FAMOUS REGIMENTS

The 11th Hussars

Other titles in this series

The Royal Fusiliers
The Royal Norfolk Regiment
The Queen's Royal Regiment (West Surrey)
The King's Royal Rifle Corps
The South Wales Borderers
The Green Howards
The Royal Berkshire Regiment
The Scots Guards
The Highland Light Infantry
The 17th/21st Lancers
The Black Watch
The Royal Hampshire Regiment
The King's Own Yorkshire Light Infantry
The Gordon Highlanders
The York and Lancaster Regiment
The Argyll and Sutherland Highlanders
The Wiltshire Regiment
The Suffolk Regiment
The 10th Royal Hussars
The Somerset Light Infantry
The Royal Flying Corps

FAMOUS REGIMENTS

Edited by
Lt-General Sir Brian Horrocks

The 11th Hussars
(Prince Albert's Own)

by
Richard Brett-Smith

Leo Cooper Ltd., London

*First published in Great Britain, 1969
by Leo Cooper Ltd,
47 Museum Street, London W.C.1
Copyright © Richard Brett-Smith 1969
Introduction Copyright © Lt-General
Sir Brian Horrocks 1969*

85052 015 0

PRINTED IN ENGLAND BY
HAZELL WATSON AND VINEY LTD
AYLESBURY, BUCKS

''Tis opportune to look back upon old times, and contemplate our forefathers. Great examples grow thin, and to be fetched from the past world. Simplicity flies away, and iniquity comes at long strides upon us.'

(Sir Thomas Browne.)

FOR
ALL THE ONES

INTRODUCTION TO THE SERIES
by Lt-General Sir Brian Horrocks

IT IS ALWAYS sad when old friends depart. In the last few years many famous old regiments have disappeared, merged into larger formations.

I suppose this is inevitable; strategy and tactics are always changing, forcing the structure of the Army to change too. But the memories of the past still linger in minds now trained to great technical proficiency and surrounded by sophisticated equipment. Nevertheless the disappearance of these well-known names as separate units marks the end of a military epoch; but we must never forget that, throughout the years, each of these regiments has carved for itself a special niche in British history. The qualities of the British character, both good and bad, which helped England to her important position in the world can be seen at work in the regiments of the old Army. To see why these regiments succeeded under Marlborough and Wellington yet failed in the American War of Independence should help us in assessing the past.

Though many Battle Honours were won during historic campaigns, the greatest contribution which our regiments have made to the British Empire is rarely mentioned: this has surely been the protection they have afforded to those indomitable British merchants who in search of fresh markets spread our influence all over the world. For some of these this involved spending many years in stinking garrisons overseas where their casualties from disease were often far greater than those suffered on active service.

The main strength of our military system has always lain

in the fact that regimental roots were planted deep into the British countryside in the shape of the Territorial Army whose battalions are also subject to the cold winds of change. This ensured the closest possible link between civilian and military worlds, and built up a unique county and family *esprit de corps* which exists in no other army in the world. A Cockney regiment, a West Country regiment and a Highland regiment differed from each other greatly, though they fought side by side in scores of battles. In spite of miserable conditions and savage discipline, a man often felt he belonged within the regiment—he shared the background and the hopes of his fellows. That was a great comfort for a soldier. Many times, at Old Comrades' gatherings, some old soldier has come up to me and said, referring to one of the World Wars, 'They were good times, Sir, weren't they?'

They were not good times at all. They were horrible times; but what these men remember and now miss was the comradeship and *esprit de corps* of the old regular regiments. These regiments, which bound men together and helped them through the pain and fear of war, deserve to be recalled.

Regimental histories are usually terribly dull, as the authors are forced to record the smallest operation and include as many names as possible. In this series we have something new. Freed from the tyranny of minute detail, the authors have sought to capture that subtle quarry, the regimental spirit. The history of each regiment is a story of a type of British life now fading away. These stories illuminate the past, and should help us to think more clearly about the military future.

The 11th Hussars
A SPECIAL INTRODUCTION
by Lt-General Sir Brian Horrocks

RICHARD BRETT-SMITH, the author, has produced one of the best regimental histories that it has ever been my privilege to read. It is so good that a personal introduction from me seems unnecessary and almost an impertinence. He has, of course, been fortunate in having some wonderful material at his disposal, because the 11th Hussars is no ordinary, run of the mill, Cavalry regiment.

What I like so much however, is that instead of concentrating on boring details of some ancient battles, which would have been all too easy in this case, as the 11th has played its part in almost all the major campaigns in which the U.K. has participated since 1715 when it was first raised by Philip Honywood, a well-known landowner in Essex, he has produced a very human story. Moreover he pulls no punches. All regiments have their ups and downs and when the 11th were down he says so, but they recovered, mainly because of their close knit family spirit which emerges from every page of his book.

The story of Old Bob is typical of what I mean. Bob was the oldest troop horse in the British Cavalry and during his twenty-five years with the regiment he was one of the few who survived the Crimean War and the Charge of the Light Brigade, and was never once struck off duty on account of sickness. He died at the ripe old age of thirty-three and was given a magnificent funeral with full military honours.

The 11th have always had a unique personality of their own; they were beloved by Queen Victoria—hence their title Prince Albert's Own; their full dress uniform, with cherry coloured overalls was the most striking in the whole British Army and during the mid-nineteenth century they must have been a perfect joy to the editors in Fleet Street, thanks to their thirty years' connection with one of the most controversial figures in British military history, James Thomas Brudenell, subsequently Earl of Cardigan. They could always be relied upon to supply front page news of a really controversial nature.

The author's description of this astonishing character makes fascinating reading and in his own words 'if the 11th could withstand, endure and even thrive under Cardigan they could withstand anything'. He was the personification of the terrible injustice of the old purchase system of promotion. Cardigan paid £40,000 to become Lieut-Colonel of the Light Dragoons, as the regiment was then called. This was a lot of money in those days so perhaps the financial gnomes of the nineteenth century might have considered he had earned his own particular niche in military history. The 11th Hussars alone in the British army sounds the Last Post every night at ten minutes before 10 p.m., instead of on the hour exactly, because that was the time Cardigan died.

When I arrived in Egypt in August 1942 the 11th Hussars equipped with armoured cars were the reconnaissance regiment of the 7th Armoured Division—the famous Desert Rats, and were by now a name to conjure with. I can say with confidence that, more than any other unit in this theatre of war they had mastered the difficult techniques of desert warfare. This involved the capacity to live for long periods on the very minimum amount of water; also considerable mechanical ability, because a breakdown which could not be repaired miles from anywhere and well behind the enemy lines meant complete disaster and a most unpleasant lingering death. They were also highly skilled in their use of the many small hills and folds in the desert for 'hull down' positions from which to observe without giving their own position away, and most important of all their information was always completely accurate. They were in fact real desert veterans and justly proud of their almost unique reputation.

There is one particular incident in connection with the 11th Hussars which I shall never forget, because it occurred on one of the few really great days of my life.

I had been sent with my small tactical headquarters plus the 7th Armoured Division and the 4th Indian Division from the 8th to the 1st British Army in order to take over command of 9th Corps and launch an attack to capture Tunis. For once the battle had developed according to plan. The two infantry divisions, 4th British and 4th Indian attacking on a comparatively narrow front, had punched a hole in the German defences through which I was able to pass the 6th and 7th Armoured Divisions down the Medjerda Valley to capture Tunis some twenty-five miles distant. It was May 7 1943 and I was standing by the roadside watching the exhilarating spectacle of these two well-trained armoured divisions sweeping down the valley in the role for which armoured divisions had been specifically designed, when suddenly a truck ground to a halt beside me, and a cheerful dusty face under the famous 11th Hussar brown beret called out 'First in again, Sir'—then roared away in a cloud of dust.

The 11th had been making a habit of being the first troops to enter many of the small towns which we had liberated during the long advance from Alamein to Tunis. As a matter of fact on this occasion it developed into a dead heat as the leading troop of the Derbyshire Yeomanry, the reconnaissance unit of the 6th Armoured Division entered Tunis by another route at the same time. But, if my memory serves me right the Derbyshire Yeomanry were commanded by an ex-11th Hussar officer, so it was almost a family affair.

When the 10th and 11th Hussars have settled down together and have become the Royal Hussars (Prince of Wales' Own) this should be a really magnificent regiment.

Acknowledgements

THIS is in no way an official or definitive history of any period in the existence of the 11th Hussars (P.A.O.). Opinions and mistakes are mine.

The book could not have been written, however, without generous help and advice from many past and present 11th Hussars and from many others outside the regiment. The list is too long to enumerate, but I thank them all sincerely.

Two fine soldiers who taught me much about the regiment in their different ways both during and after the war were—and happily remain, although both have long since retired from professional soldiering—Lieut-Colonel John Turnbull, M.C., and Major Toby Horsford, M.C. I trust that they will forgive me for mentioning them here.

Many people have been kind in producing or suggesting photographs and illustrations, only a few of which can appear in the book. So too of written sources of information. Previous regimental histories have been invaluable, and in the context of the Crimea it would be churlish not to mention *The Reason Why*, by Mrs. Cecil Woodham-Smith, and *The Destruction of Lord Raglan*, by Christopher Hibbert.

Special thanks to: Colonel Sir William Honywood, Bt., M.C., and Filmer Honywood, Esq.; Major T. I. Pitman, M.C.; Colonel H. N. H. Wild, O.B.E.; Lieut.-Colonel W. B. R. Neave-Hill, Narrator, Ministry of Defence (Central and Army) Library; the Bodleian; the Imperial War Museum; the Royal United Service Institution; Cavalry Club; the Victoria and Albert Museum; the Ringling Museum of Art (Sarasota, Florida) and the Royal Academy.

R. B-S.

Field Marshal Sir Philip Honywood, founder of the Regiment.

Royal Academy

General Philip Honywood, the nephew of the founder, painted by Gainsborough.

Chapter

I

'Mr Secretary Pulteney desires to see you at his office . . .'

IN June 1694, some six months before the death of his sovereign Queen Mary, Philip Honywood, younger son of a squire from Kent with property in Essex, was gazetted Ensign of Foot. His subsequent military career was to be a spectacular one, in setbacks as well as in successes.

It was not long before Honywood saw action, for in 1695 he was present at the recapture of Namur by his King, William III, of whom he was a faithful supporter. But William and Mary had not been blessed with children, so that when the King died in 1702 Mary's sister Anne became Queen, who, astonishingly even for those days, had lost the last of all her many children through illness by 1700. The succession to the English throne thus devolved upon Sophia, Electress of Hanover, the nearest Protestant descendant of James I, and her son, destined to become George I.

But no such seemingly orderly sequence of events was clear to young officers such as Honywood, especially when the complex and lengthy War of the Spanish Succession broke out in 1701, with all its undertones of international intrigue and double-dealing. These were reflected in England's domestic politics, not least because Louis XIV of France proclaimed James II's son James Edward, the 'Old Pretender', rightful King of Great Britain. In England the Whigs, wholehearted supporters of the Glorious Revolution of 1687–88, and thus of William of Orange, the

Dutch alliance and the Hanoverian succession, were the war party against the French on sea and land; the Jacobites, prepared to accept Anne but not the Hanoverians, remained quiescent and mainly unvocal until the Union with Scotland in 1707, which aroused keen antipathies. Thus many Jacobites during the years of some of the finest victories of one of England's greatest soldiers, John Churchill, first Duke of Marlborough, would support the non-Jacobite Tories, the party of the country landowners, who accepted the Revolution while anxious to curtail its results, and who were loth to finance an expensive war against the French, except at sea, seeing little chance of removing Louis XIV's grandson Philip from the Spanish throne.

Marlborough, by background and inclination a Tory of the kind Anne preferred, had not been above intriguing with James II after the Revolution and after James's death flirting with the Old Pretender, whose cause, let it be said, attracted much emotional support in England. Marlborough after all had been a close supporter and confidant of James II before deserting him and going over to William III, though always a firm Protestant. He was more or less driven into the Whig camp during Anne's reign because only the Whigs supported the large expenditure necessary for the Spanish war (against the French).

Although the Captain-General was still immensely powerful in 1707–08, his cause and financial reputation were gradually eroded by his political enemies on the Tory side, by popular unrest with the unending war, and by the alienation of his Duchess [Thackeray's 'That horrible Duchess'] from Queen Anne, over whom (as over her husband), she had been for thirty years the dominating influence. Thus in 1810 the Whigs were finally ousted by the Tories under Harley and St. John (the latter sympathetic to the Jacobites), and Marlborough was dismissed with odium.

Philip Honywood, coming from a landowning family

similar to John Churchill's in Dorset, was a natural supporter of Marlborough; there can be little doubt that he was, however, more ardent and loyal a one than many of his contemporaries. He had seen two campaigns in Brabant under the Duke, and early in 1704 was rewarded for his excellent conduct in the Spanish war by being made Lieut-Colonel of Wade's (later the 33rd) Foot when well under the age of thirty; by September 1710 he was a Brigadier-General, an achievement even in the age of purchase.

Very soon, however, this heady record of promotion was to come to nothing, with Honywood's commander Marlborough either unable or unwilling to save his devoted junior. The Rev. Jonathan Swift, writing from London on December 13, 1710, in his Journal to Stella, presents a civilian's comment on the consequence of the Whigs' discomfiture:

> 'You hear the havock making in the army: [Lieut-General] Meredyth, [Major-General] Maccartney, and Colonel (*sic*) Honywood, are obliged to sell their commands at half value, and leave the army, for drinking Destruction to the present Ministry, and dressing up a hat on a stick, and calling it Harley; then drinking a glass with one hand, and discharging a pistol with the other at the maukin, wishing it were Harley himself: and a hundred other such pretty tricks, as inflaming their soldiers, and foreign Ministers against the late changes at Court.'

Apparently Stella (Esther Johnson) contested Swift's account, for the following January 31 he is moved to retort:

> 'Your intelligence that the story is false about the officers forced to sell, is admirable. You may see them all three here every day, no more in the army than you.'

These bold anti-Tory sentiments expressed by three Generals, all of them Marlborough's men, might well have cost them their careers finally; for if not exactly cashiered because of them, they suffered the next best thing. Honywood's loyalty was admirable enough, his choice of companions, at least in one case, rash. For in 1712 Maccartney (and possibly Meredith too) was involved in the notorious two- or three-sided duel in which the Tory Duke of Hamilton and the Whig Lord Mohun died, the former according to Swift and Thackeray (in *Henry Esmond*) stabbed by Maccartney when already wounded by Mohun, although the murdering blow may have been struck by one of Mohun's servants. Mohun was an infamous rake and duellist whose loss to the Whigs and Marlborough can have been regretted little by decent people, and Meredith and Maccartney were his especial cronies. They still contrived to work their passages.

Honywood, like them, though uninvolved in the shady 'duel', was *persona non grata* until after Queen Anne's death in 1714. General Meredith was also reinstated ultimately, but not before he had lost most of his money through mismanaged ventures on the stock market. Maccartney, 'That villain' as the Tory Swift calls him, a veteran of Malplaquet and other battles, escaped to Holland after the Hamilton affair despite a Crown bounty of £500 for his apprehension, and only returned to England after George I's accession. Nevertheless, in June 1716 he was tried for murder, found guilty as an accessory, and 'burnt' in the hand, despite the discrediting of the Duke's cousin and second, Colonel Hamilton, Scots Guards, who gave evidence against Maccartney. Ironically enough Colonel Hamilton was forced to sell out and leave the country, whereas within a month Maccartney was promoted to Lieut-General and given an appointment. (He reviewed Honywood's new regiment at Stanford in 1718 and died in 1730.)

With George I narrowly but safely established on the throne (his mother having died shortly before Queen Anne), the Protestant House of Hanover began its long rule and the

Whigs came back in favour. None could foresee, however, how long the Hanoverians would last, and many leading Tories, fearing a Whig monopoly of power, began or resumed negotiations with James Edward. The Catholic line of the Stuarts was officially excluded, but by 1715 Jacobite riots flared in England, and in France Louis XIV promised arms and money to the exiled Pretender.

None too soon George I took precautionary action by augmenting the Army by eight regiments of Foot and thirteen of Dragoons, Parliament approving this measure in July, 1715. Less than two months later the Earl of Mar raised the Pretender's standard at Kirkmichael, the clans flocked to the Chevalier's White Cockade, Scotland was aflame and the North of England threatened. For Honywood it was opportune. His loyalty to the House of Hanover was rewarded. Although an infantryman, a man of his background would have been brought up with horses, and his was one of the names selected by the Secretary of State, William Pulteney, to be Colonel of a regiment of Dragoons. In those days the Colonel, thought he might—or might not —leave command in the field to his Lieut-Colonel (being himself likely to be commanding a senior formation) was virtually omnipotent over his regiment; it was he who dealt direct with his superiors and the government concerning his regimental equipment, uniforms, pay, postings, ranks, and welfare, his regiment owing to and transacting everything through him alone. Obviously, a very great deal depended upon the calibre of a regimental Colonel, and Honywood's Dragoons must have counted themselves fortunate. Of the other cavalry regiments raised simultaneously, the Dragoons of Major-Generals Pepper and Wynne later became respectively the 8th Hussars and 9th Lancers, those of Brigadier-Generals Gore, Bowles, Munden and Dormer respectively the 10th Hussars, 12th Lancers, 13th and 14th Hussars.

Honywood had little time to raise and train his regiment;

the warrant for their raising is dated July 22, 1715, Essex and adjoining counties being the recruiting area, with Chelmsford as headquarters, and Braintree, Bockin and Whitham as other quarters. Time was indeed short, for by November Honywood's Dragoons had gone north to Cheshire and Lancashire as part of the expedition commanded by Major-General Wills against the Old Pretender's forces.

Pulteney had laid down that each new regiment of Dragoons should consist of six Troops, each of one sergeant, six corporals, one drummer, one hautboy, and twenty-eight privates; Honywood hand-picked seventeen officers, so that when a further sixty other ranks were added to the regiment in late October his total strength was exactly 300 all ranks. Their uniform was scarlet coats turned up with buff, buff waistcoats and breeches, jacked leather boots, and tricorne cocked hats bound with silver lace; buff again was the colour of the horse furniture, and as in the rest of George I's Army their guidons carried the House of Hanover's badge, the white horse of Brunswick.

As first Lieut-Colonel of his regiment (and Captain of a Troop) Honywood chose another experienced infantryman, Archibald Hamilton, who had fought under William III at the Battle of the Boyne, and later had campaigned in Flanders, Portugal and Spain. Hamilton was to become a Lieut-General. So too was Humphry Bland, Honywood's Major, another veteran destined in years to come to be C-in-C Scotland after Culloden, and to find even wider fame as the author of perhaps the best-known English military textbook of the eighteenth century, *Bland's Military Discipline*. One other of the original officers must be mentioned, Cornet William Gardner, who was to achieve a record of continuous service in the regiment second to none and probably unique for an officer in that age of frequent purchases and cross-postings—forty-five years, nearly seven of them in command.

So, raw but doubtless crash-trained and disciplined by able campaigners such as Bland and Hamilton, early in November 1715 Honywood's Dragoons went to war, mounted if old tradition be right on greys. They moved via Nottingham to Chester, and thence to Manchester and Wigan, where Wills was gathering his forces before attacking Preston, which the Jacobites had taken. Lieut-General Carpenter advancing with another force from Durham planned to take the rebels in the flank in Lancashire.

The rebels, numbering some 1,500 foot and horse, mostly Scottish but with a sizeable contingent of northern English Jacobites, were commanded by 'General' Forster and Lord Kenmure; they had already occupied Penrith, Appleby, Kirkby Lonsdale and Lancaster. Their defence of Preston, however, from which they had already ousted Stanhope's Dragoons (another unblooded new regiment) and a force of militia without firing a shot, appears to have been singularly inept. There was plenty of time to organise excellent positions, but Mr Forster spent most of it in bed, with Lord Kenmure left to give out orders. Inexplicably no one decided to man the River Ribble on the town's outskirts, a natural defensive position, or the road from the bridge across which Wills's force was almost bound to come to the town, a difficult road with steep banks on either side, upon which Oliver Cromwell in 1648 had met spirited resistance from the Royalists. Instead the rebels relied upon a makeshift and incomplete series of barricades within Preston itself, supported by cannon. These were detected easily by General Wills who had made a personal reconnaissance of the town.

Wills had plenty of cavalry, but only three regiments of foot, upon which the brunt of the attack would obviously fall, and no artillery. So a number of troops from the Dragoon regiments were dismounted, Honywood's Dragoons contributing fifty men to the force that stormed the Wigan Avenue. This attack was directed by Brigadier-

General Honywood himself, who was commanding his own and Wynne's regiments and some of the foot. Wills had formed two other cavalry brigades under Brigadier-Generals Munden and Dormer.

The fighting was fierce, scrappy, and inconclusive on November 12, 1715, mainly due to the attackers' lack of cannon.

The rebels' second and stronger line of barricades managed to hold out after the first had fallen, and for the inexperienced new cavalry, condemned to what amounted more or less to street fighting and cordoning the approaches to the town, it must have been a frustrating day. In casualties the rebels came off best, losing three officers and seventeen men killed, and twenty-five wounded, including Major Preston, an excellent officer of Lord Forrester's Foot (the Cameronians), who had made a very gallant counter-attack and lost sixty or seventy lives including his own in the process.

Brigadier-General Honywood, who showed 'signal valour and judgement' according to Cannon, was wounded in the shoulder, Major Bland in the arm, and five privates and twelve horses of the regiment were wounded. Its contribution had been small but spirited.

General Carpenter and his three Dragoon regiments had not arrived on the 12th, but the following morning, a Sunday, they were observed approaching, from a church steeple. At this news 'General' Forster at once beat a parley (without having consulted his senior officers), and early on the 14th 1,468 Jacobites, 463 of them English, laid down their arms and surrendered to Generals Wills and Carpenter when they entered Preston. Many of the rebels were men of high rank and position, both Scots and English, and a number, including Lord Kenmure and the Earl of Derwentwater, were later executed. Mr Forster was to escape from the Tower of London.

The Hanoverian victory at Preston effectively damped

out Jacobite hopes in England; and even in Scotland, where the Pretender had finally landed in December, 1715, the cause was becoming Laodicean. Early in February 1716 James Edward, his army dispersed, had to flee his country in a fishing smack and return to France. 'The Fifteen' was over.

*   *   *   *   *   *

Honywood's men were largely raw countrymen and yeomen from the south of England, who had had to be whipped into shape rapidly by a few professionals. Their ranks, however, did contain a few genuine old soldiers, and one who stands out entertainingly is Donald McBane, who enlisted, by no means for the first time, when well over fifty and was made up to sergeant in the regiment. McBane carried the standard at Preston, but was forced to seek his discharge next year because the harsh winter spent by the regiment at Bolton caused his old leg wounds to break out. Here was a real 'old sweat', who had served in regular and irregular regiments all over the place, and who had a deserved reputation as one of the finest swordsmen in the British Army. But his sort of soldiering career was by no means unusual in the latter part of the seventeenth and in the eighteenth century. In fact it is worth quoting from the address in verse to the reader that prefaces his curious hotch-potch of a volume, *The Expert Sword-Man's Companion* (Glasgow, 1728, now an extremely rare book since it contained instructional diagrams which would have been torn out or dog-eared by use), to give some idea of what could happen to the professional soldier of those times.

> 'Time was, the Author Travel'd far and near,
> Under the notion of a Musquetier;
> And shortly after to a Pike-Man rose,
> Plac'd in the Fore-front to offend our Foes.
> Soon after for the Space of Twenty Years,
> Was I one of the Royal Granadiers:

Inroll'd in Lord George Hamilton's Command,
The Hope and Honour of our Native Land.
In Sixteen Battles Foughten, I have been,
And Fifty-two great Sieges I have seen.
Five-score and Sixteen Times I did Advance,
In Flanders, Holland, Germany and France.
My Fourth Course was a Sergeant of Dragoons,
Well known at Preston, and at other Towns.
And lastly I'm Fort Williams Cannonier,
Thanks be to GOD, my En'mies I don't Fear.'

Sergeant McBane, who must have been born in about 1663, had served under Marlborough amongst other commanders, and was a noted duellist. When he left for a time the Army in 1712 he got married and kept an ale-house and a fencing school in London where he 'lived very well' and contrived to fight thirty-seven prizes in Bear Garden. He was also something of a gunnery expert, and in later life at any rate inclined to religion. This did not prevent his being incorrigible at a fairly advanced age regarding his favourite pursuit, for he ends his book:

'In 1726, I Fought a Clean Young Man at Edinburgh, I gave him Seven Wounds, and broke his Arm with the Fauchion, this I did at the Request of several Noblemen and Gentlemen. But now being Sixty-three Years of Age, resolves never to Fight any more, but to Repent for my former Wickedness.'

For nearly thirty years between the two big Jacobite rebellions the regiment was engaged on routine peacetime duties in England, mostly in the Midlands and Home Counties, sometimes in the South and West Country, sometimes in the North, and once in Scotland. Cavalry regiments went annually to pasture quarters between April and September, for cavalry barracks were not built until 1794. Most of the duties during this long time were unromantic, and the dragoon of those days was neither widely admired nor esteemed. Yet in this apparently dull period may be seen the evolution of the naïve countryman of 'The Fifteen' into the rough if unloved professional soldier under Cumberland of 'The Forty-Five.'

By 1718 the regiment's establishment was down to only 207 all ranks (whose estimated cost was £11,226 15s. 10d. yearly) and it remained very low for some years. Indeed in 1726 Honywood got sharply rebuked after an inspection by his old commander Sir Charles Wills, now a Lieut-General, because he was one sergeant and a dozen men under strength. He was told from Whitehall:

> 'H.M. is much surprised, and hath commanded me to let you know that he expects you to be more careful in future. . . .'

In 1727, when George II succeeded, England had already embarked upon her golden and stable Augustan age, but there was a fear of a German war that year so the regiment was readied for foreign service and brought up to 552 men; but it never embarked and in Scotland next year was down to 309 men. (Its normal strength between 1723–1728 was between 300–400 all ranks.)

In 1732, after nearly seventeen years' continuous control, Philip Honywood removed to the 3rd Dragoons and the Colonelcy was given by George II to Major-General Lord Mark Kerr, Colonel of the 13th Foot. So after the prevailing fashion 'Honywood's Dragoons' now became 'Kerr's Dragoons'. In 1733 the King expressed himself very pleased with their appearance and discipline at a review on Hounslow Heath, and next year every dragoon regiment was given another troop, consisting of three officers and fifty-eight other ranks, bringing Kerr's Dragoons up to 370 all ranks. In 1739 because of the War with Spain over Jenkins's Ear it was up to 435 men.

Most of their duties during this happy period were in aid of the civil power, conducting prisoners or valuables, coping with occasional riots in towns (for some reason weavers were especially prone to them), helping to quell the odd mutiny, and in particular constantly putting down 'owling' —the illegal export of wool and sheep—and smuggling. Soldiers became familiar with stations such as Preston, Ormskirk, Newcastle, Warwick, Coventry, Worcester, Lichfield, Reading, Lewes, Norwich, Ipswich, Canterbury, Wakefield and Hounslow. Occasionally a draft of men and horses had to be provided for Flanders or overseas, but in the main it was a placid time for the soldiers. Not always for the civilians, however, who made their usual complaints of rowdiness and poaching game. Inn-keepers and landlords were liable to be fair game, too, for nearly always troops or detachments were split up among local inns for billeting if municipal buildings were not available, or often if they were. An anonymous essayist in the *Gentleman's Magazine* of 1740, when the War with Spain was on, is particularly resentful of the hazards of inn-keeping and the brutal manners of the soldiery, and it is a matter for regret that the following account may well refer to some of Lord Mark Kerr's Dragoons (who were in camp near Hounslow that spring!).

Every soldier quartered on an inn-keeper, 'The Craftsman' reckoned, cost him nearly as much as he cost the government,

> 'not to mention the insolent Behaviour, Incroachments, Waste, Debauchery, and leud Examples of such profligate and almost lawless Guests.'

'I once happened to lye at a Country Inn,' the writer goes on plaintively,

> 'not very far from Hounslow Heath, where there was to be a Review some Days afterwards. When I came there, I found no less than twelve Dragoons, and toward the Evening as many more came rattling into the Yard, took up all the best Stabling, and litter'd their Horses up to their Bellies with the best Straw the House could afford.

Nay, they made the Ostlers, and other Servants rub down their Horses which is not their Business, but the Master of the House told me that he dar'd not disoblige them, (though the first Nobleman in the Land should call in there;) for if he did, they would be revenged upon him by wasting his Hay, and otherwise doing him all the Mischief that lay in their power. He was then to provide them with an hot Supper of the best Meat that could be bought; for their nice Stomachs could not digest cold Victuals, nor such ordinary Provisions as often serve the Family at a Country Inn. The next Morning the Master of the House desired me to walk into the Kitchen, and see the Gentlemen, as he call'd them, at Breakfast; which consisted of a great Piece of cold Roast-Beef, Shoulder of Mutton, two or three Peck-Loaves, and a Cheshire Cheese, the greatest Part of which was devour'd in an Instant. They were then to have an hot Dinner intirely to themselves, for they would not suffer a Slice to be cut off for any Customer; and this the poor Man told me would be his Case, from Day to Day, as long as he was plagued with their Company.'

Clearly one could live well enough off the land even at home two hundred years ago, without benefit of Public Relations.

Chapter 2

Honywood, who had commanded a brigade in Spain in 1719, had been promoted Major-General in 1726, and in 1735 both he and Lord Mark Kerr, whose military career closely paralleled his own, were made Lieut-General; in 1743 they were both full Generals (of Horse and Foot respectively), Honywood having commanded the army in Flanders until Field Marshal the Earl of Stair's arrival, and then led the cavalry of the front line with great distinction at Dettingen: where his nephew, also Philip Honywood, who had followed his uncle from the regiment to the 3rd Dragoons, showed notable gallantry, was wounded five times, plundered, stripped and left for dead by the French, but recovered to fight them again two years later at Fontenoy. The younger Honywood was also destined to become a full General. Kerr was wounded in 1707 in a desperate action at Almanza leading his own regiment of Foot which was practically wiped out, and also commanded a brigade at the siege and capture of Vigo in 1719, in which the elder Honywood took the leading part.

In May, 1743, General Honywood was created Knight of the Bath, for his services in Flanders and Germany, being invested that August by George II 'at the Head of the Army'. In December he must have been gratified when his nephew was made a Lieut-Colonel in the 3rd Dragoons, though the younger Honywood never actually commanded Kerr's Dragoons after Lieut-Colonel (later General) Hugh Warburton.

It was now high time for the regiment to prove in action

what had been the fruits of long and placid years of training, since it had been denied service abroad. The opportunity came at last; in July 1745 the 'Young Pretender', Charles Edward, landed in Scotland from France and in August, having proclaimed his father James VIII of Scotland and III of England, raised the Stuart standard at a gathering of the Clans at Glenfinnan. He entered Edinburgh without opposition on September 17, and four days later routed the King's forces under Sir John Cope at nearby Preston Pans. When the Hanoverians reached Berwick in disorder, Lord Mark Kerr, who appears to have been acting Governor there in succession to Honywood, remarked acidly that Cope must be the first general to bring the news of his own defeat. Kerr was a character, with very strict notions of honour, good-breeding, and etiquette. Soldierly in appearance, though finical in dress, his over-punctiliousness led him into a good many duels, which he made a habit of winning. But we are told that he was 'naturally of a good temper' and 'inoffensive unless provoked', and that 'he commanded respect wherever he went, for none dared to laugh at his singularities' (*Douglas's Peerage by Wood*, quoted by Cannon).

Encouraged by his success, and having augmented his army from some 2,000 to 7,000 men, the Prince moved daringly with two columns (the other under Lord George Murray) into England, hopeful of a mass rising in his support. But he was disappointed by a lukewarm reception, even in Jacobite Lancashire, and though Carlisle, Preston, Wigan and Manchester gave in to him, and by clever marches he put himself between the Duke of Cumberland's army and entered Derby on December 4, that was the limit of his advance. French help largely failed to materialise, there was no English uprising on his behalf, and he was greatly outnumbered, facing in all a force of about 30,000 mostly regular troops split into four divisions. Reluctantly acceding to his chieftains, Charles had to retrace his steps, closely pursued by Cumberland's enlarged 3rd Division of fifteen battalions, five of which were horse, amongst them Kerr's Dragoons.

On December 18 Cumberland's cavalry skirmishers overtook the Pretender's rearguard at Clifton Muir, near Penrith, but in a confused dismounted attack by night, in which one troop of Kerr's took part, victory went to the defending Highlanders, who then resumed their withdrawal unharassed. By the end of the year Cumberland had reduced Carlisle, but on January 17, 1746, General Hawley was defeated severely by Charles's forces at Falkirk. Nevertheless, the Pretender's cause was only really strong in the Highlands proper, and Cumberland pressed on to Edinburgh, Stirling, Crieff, Forfar, and Aberdeen.

Lieut-Colonel the Earl of Ancram (afterwards 5th Marquess of Lothian), who had started his military career as a Cornet in Kerr's Dragoons, was now commanding them in the field after returning from the 1st Foot (Grenadier) Guards, and early in March 100 of his men took unopposed and destroyed a large dump of Spanish arms and powder at Corgarff Castle, near Aberdeen, in the heart of rebel country. In a letter to the future Prime Minister, the Duke of Newcastle, at Whitehall, Cumberland noted of Ancram that:

> 'during the whole expedition he has behaved with the greatest prudence and caution and much like an Officer.'

As A.D.C. to Cumberland he had been severely wounded at Fontenoy the year before by a musket-ball in the head.

A 'parade state' of March 28 shows that the regiment mustered 267 soldiers, of whom eighteen were sick, with another seventy-eight sick or wounded in hospital; 283 horses, all apparently 'black', were fit, twenty-six sick, and fifty-four had been left behind. By direction light dragoons were restricted to a height of five feet eight inches, and they were lightly mounted with a limit of fifteen hands.

By April 13 Cumberland had crossed the Spey, and two days later had reached Nairn by way of Elgin and Forres, inflicting minor losses on the retreating Jacobites. On April 16, considerably to Cumberland's surprise, the rebels finally stood and risked a decisive battle at Culloden, some six miles east of Inverness. It was a disaster for them, bravely as they fought. Wild charges by their infantry availed nothing against the well-trained and disciplined regular infantry, the rebels' cannon was 'extremely ill served and ill pointed' (Cumberland), and Charles was short of cavalry compared to his opponents. It was in fact cavalry charges from both wings that met in the centre

Army Museums Ogilby Trust

William John, Earl of Ancram, and later 5th Marquess of Lothian. Colonel, 11th Dragoons.

which set the seal upon the rebels' defeat, and Ancram, ordered to lead the pursuit of the broken enemy, inflicted terrible casualties for several miles across Culloden Moor. So for the first time the regiment fulfilled its proper function and fought a major cavalry action. Some Scottish writers have accused the King's horsemen of needless brutality amounting to cowardice at Culloden. Certainly the slaughter was ruthless, but Ancram, in the words of John Prebble, was 'a chivalrous and compassionate man'. Cumberland first estimated the Jacobites' losses at 2,000, but later upped it by as many again after hearing their own accounts. Few wounded got away ('The Monro's have knocked on the head fifty of the Rebels in their flight' Cumberland related with satisfaction), and all Charles's artillery, ammunition and baggage were captured. Over 200 French prisoners were taken.

On the King's side (which included not a few 'loyal' Highlanders) the infantry bore the brunt, but the loss in killed and wounded of barely over 300 was remarkably light. Kerr's Dragoons, which lost only three men killed, three wounded, and some twenty horses wounded, were congratulated by Cumberland for their zeal and gallantry. They had captured Lord Balmerino, who was later executed.

Prince Charles Edward himself headed south-west for the country of his fiercest supporters, the Camerons, and for months while his fugitive soldiers were relentlessly hunted down or hid starving in wild country he miraculously evaded capture. Despite the reward of £30,000, a huge sum then, offered for his body dead or alive, no one of his people betrayed him, and despite hairbreadth escapes from Cumberland's men, who watched like hawks even tiny fishing ports, the Pretender finally escaped to France in September 1716, having sailed right through the English fleet in fog. The last Stuart bid for the English throne was over.

★ ★ ★ ★ ★ ★

By 1747 the regiment was back in England, and as

hostilities in Europe came to an end the following year, its establishment was reduced in 1749 to 285 men. In 1751 George II, a keen soldier, laid down definite dress regulations for his regiments. The uniform of Kerr's Dragoons remained basically scarlet and buff in colour, but for the first time we meet the characters 'XI D' embroidered on a red ground on the buff saddle-cloths, in silver on buff ground on the crimson King's Guidon, and in silver on crimson ground on the buff second and third, or regimental, guidons.

In 1752 came two great losses: first the Colonel of nearly twenty years, Lord Mark Kerr, died, his grand-nephew Ancram succeeding him; then in June came the death of Sir Philip Honywood, the oldest General of Horse in the Army. Honours had come thick upon him in his latter years —he had reviewed the Forces, been given another Colonelcy, been made Governor of Portsmouth and an A.D.C. to the King, and best of all, if the *Gentleman's Magazine* for January, 1751, is accurate, been 'appointed Field Marshal, in room of Gen. Wade, dec.'.

How delighted the old man must have been, who had lived to see his own dragoons distinguish themselves once more in battle. Although it was to be generations before his regiment acquired its peculiar motto, he at least through countless ups and downs had proved himself not only a brave and experienced campaigner but 'true and faithful' to his troops as to his Sovereigns.

In the quiet years between 1746 and 1758 there is little to record. As has been seen, the size of the regiment depended considerably on whether or not there was a Continental war in progress; thus in 1749 it was down to 285 men, in 1755 up to 357, because of the North American war against France, and in 1756, when Lieut-General Campbell inspected Ancram's Dragoons—'an extreme fine regiment'—they turned out 380 mounted men. From 1755 on the number '11' is constantly associated with the regiment, known now usually as 'Lord Ancram's Dragoons' but sometimes as 'the 11th Regiment of Dragoons'.

The most interesting event of this period for the British cavalry was the introduction in 1756 into eleven regiments of a Light Troop of sixty men. At last some of the lessons learned in warfare on the Continent were being digested and put to good use, for in cavalry training, tactics and deployment in warfare the English had much to learn from leaders such as Frederick the Great, Marshal Saxe, Prince Eugen and Count von Daun. Before these Light Troops were first reduced in 1763 and finally disbanded altogether in 1779 (by which time whole regiments of light dragoons had been formed, so the lessons were not lost), nine of them had performed useful service in expeditions in 1758 against St Malo and Cherbourg, none more so than that of the 11th under Captain William Lindsay, unfortunately killed near Cherbourg.

Army Museums Ogilby Trust

A trooper of the Light Horse, which was created in 1756 to act as Hussars. From a drawing dated 1762.

Contemporary journals had already christened these light troops 'Hussars', after the Hungarian light cavalry, and their aptitude for fighting reconnaissance justifies their being called the direct ancestors of the armoured car troops of the 11th Hussars some 180 years later. Someone at last had bothered to place emphasis in the right places, on speed, mobility and adaptation of the training and tactics of the best cavalry in Europe—that of the Prussians and Austrians. By Warrant of April 14, 1756, each Light Troop consisted of:

> 'One Captain, one Lieut., one Cornet, one quarter Master, two Sergeants, three Corporals, two Drummers, and 60 Light Dragoons, including a Farrier.'

The same Warrant directed:

> 'The Corporals and private men to be from 5 feet $6\frac{1}{2}$ inches to 5 feet 8 inches high and light active men. The size of the horses 14-3 and not under, well turned nimble road horses, as nigh the colour of the Regiment as can be got.'
> (Since about 1753 this had been dark brown in Ancram's Dragoons, not always an easy colour to get enough of.)

Light troopers were equipped with a carbine with ring and bar measuring four feet three inches, a bayonet seventeen inches long, a pistol ten inches in the barrel and of carbine bore, and a straight cutting sword thirty-four inches long. All their accoutrements were of tanned leather, and they used small, almost jockey's saddles and light bits and bridles. Their uniform was like the rest of their regiment, except that they wore jacked leather caps and light jockey boots with stiff tops. They were certainly not under-equipped, for on the right of the saddle they carried pistol holsters and on the left a churn containing a spade and a felling axe or woodman's bill. The War Office allowed £5 16s. 9d. per each N.C.O. and private for accoutrements and horse furniture, £3 levy money, and £12 per horse. A private soldier's pay at the time was only 6*d*. a day (before

1718 it had been 8*d*.), which it was to remain until 1797, when it went up to 1*s*. a day—at which it remained constant, unbelievably, for over a century. After messing and other stoppages only a penny or two was left.

Ancram's Light Troop became the first of the Eleventh to see service and action abroad, for the Seven Years' War, which had started in North America in 1755, spread to France and Germany on land and in all directions at sea. Thus in the summer of 1758 the Light Troop took part in three large expeditions against St Malo (twice) and Cherbourg. In the first St Malo raid, in which Ancram commanded a division, ninety French ships and a large amount of stores and magazines were burned by the British dragoons, though an attack on the town itself could not be mounted. In August, however, Cherbourg was taken by a seaborne landing and its fortifications, harbour and ship-

Army Museums Ogilby Trust

'11th Dragoons fighting the French.' From a painting thought to be by David Morier. Reproduced by gracious permission of Her Majesty the Queen.

ping fired with great damage resulting. Unfortunately the work of destruction

> 'was much retarded by the drunkenness of certain men who, having come upon wine vaults, proceeded to take advantage of their discovery',

so *The Historical Records of the Eleventh Hussars* relates. This lapse, from which probably some of Ancram's Dragoons cannot be exculpated, was not unknown nearly 200 years later among American soldiers at Cherbourg in 1944. In 1758, however, the penalty in many cases was a death sentence. 'Looters were also murdered by peasants, while marauding in the district.' Ancram's Light Troop did distinguish itself in covering operations and skirmishes as far inland as the town of Valognes. The second St Malo raid in September was not so successful, as the French mustered a substantial force which eventually compelled a sort of miniature Dunkirk on the beach. The British rearguard, mainly of Grenadiers including the 1st Foot Guards, was badly mauled and many men had to swim for their lives to the waiting ships. The 11th detachments which had again been skirmishing inland, lost half its horses and a third of its men in this embarkation, and altogether about 700 British troops were killed, drowned, or taken prisoner, the French casualties being nearly as high.

Back in England in 1759 the 11th was required to find drafts for other regiments serving in Germany, and these replacements had to be men at least five feet eight inches tall and between the ages of twenty and forty. At an inspection this year the great scarcity of dark brown horses is remarked—which doubtless led in 1765 to Royal assent that in future black horses could be purchased instead. Interestingly in 1759 strict orders were in force for the rejection of undesirable officers, witness a letter from Barrington, Secretary of War, thanking Ancram for having objected to some officers in line with an order proposed by

Barrington 'as a security against improper persons slipping into the Army'. Cumberland, embarrassingly compelled to capitulate to the French at Kloster-Zeven in 1757 after the defeat of his Hessians and Prussians at Hastenbeck, had had the same trouble.

In April 1760, the year when George III came to the throne, the regiment went for the first time to Germany, where Frederick the Great was conducting a brilliant series of campaigns against the French and their allies. This was to be its first prolonged dose of foreign campaigning, for it was nearly three years before the 11th came home, after fighting many actions, often alongside their German allies. Under Lieut-Colonel William Gardner, who must have been getting on for seventy and whose last year of

Army Museums Ogilby Trust

A farrier of the 11th Dragoons. The farrier's task was to kill any injured horses and remove two of their hooves as evidence; this perhaps accounts for the disconsolate attitude of both man and horse.

forty-five years' continuous service this was, they joined the army of Prince Ferdinand of Brunswick and took part in the successful battle of Warburg, where though in Ferdinand's words 'All the British cavalry performed prodigies of valour' their casualties were minimal. The splendid charge which routed the French horse across the Diemel in complete disorder was led by the Marquess of Granby, riding at the head of his own regiment, the Blues; Lord Granby lost his hat and charged literally bald-headed, his pate shining in the early morning sun!

Warburg provided the first battle honour that the regiment had won. During the next two years there was sporadic heavy fighting in the plains and forests of Hanover and Hesse, particularly in the Cassel-Göttingen area, the French sustaining 5,000 casualties at Kirch Denkern between the Lippe and Asse rivers; equally there were long periods of routine patrolling or static defence, the weather was often bad, and the winters very bitter. During that of 1761 the regiment withdrew to cantonments in East Friesland. By this time Colonel Gardner honourably decided that he had served long enough, and retired, collecting the very substantial sum for those days of £3,675 from his successor, Major John Bradford, and three other officers who thus bought promotion according to the prevalent custom. On June 24, 1762 Prince Ferdinand took the French by surprise at Grebenstein. Abandoning their tents they streamed back towards Cassel, but lost 4,000 men in the woods near Wilhelmstal. The 11th, still under Granby's generalship, took part in this action and in the siege of Cassel that followed.

After Cassel and Göttingen had fallen to Ferdinand in July the war petered out, and in February 1763 was officially ended by the 'Peace of Paris' and the 'Treaty of Hubertusburg'. The 11th, heartened by the commendations of Ferdinand and Granby, and the thanks accorded the whole Army by the British Parliament, left Germany and marched through Holland to Willemstad on the sea. From this small port they reached England in March, and the Light Troop was now disbanded, its place being taken by men from individual troops mounted as light cavalry.

The next thirty years were spent on routine duties in England and Scotland, closely resembling the period between the two Jacobite rebellions. The regiment took no part in the American Wars, though it did provide drafts for other regiments, notably the new 20th Light Dragoons in 1779. In 1775 Major-General James Johnston had succeeded the Marquis of Lothian (Ancram) as Colonel, but rather than 'Johnston's Dragoons' the regiment was now more usually known as '11th Light Dragoons', a title which became official in 1783. As a result the standard height for both men and horses was lowered, equipment was made lighter, and the helmet replaced the cocked hat. In 1784 the coat, which had always been scarlet, was changed to blue, with buff facings as before.

During these years the standard of mounts (from 1765 black again instead of brown) is often criticised, sometimes severely. The strength of the regiment also varied considerably; thus in 1769 only 186 mounted men turned out for the Inspector-General, in 1781 221, and in 1782 regimental strength was probably about 250 men.

One should recall that a regiment of dragoons on home service in the eighteenth century seldom resembled a regiment. Inevitably it was split up into small detachments, usually quartered at local taverns and inns. There were still no cavalry barracks, so for a good part of the year horses were put out to grass. There was a fairly high turnover of men, and especially of officers, owing to the cash promotion system, though the impoverished might stay for years in the same rank with the same regiment. There were frequent though usually small drafts to other regiments for overseas service. In 1788, for example, all 'heavy' men in the 11th were posted to 'Heavy Dragoons'.

The frequent slight strength of the 11th over this period may be explained partly by the fact that one Commanding Officer at least, Lieut-Colonel (later General) Ralph Dundas, no doubt partly because of inspection criticisms, 'cast' a considerable number of men and horses. This got him into trouble with the War Office, with whom he had many acrimonious arguments on paper, from which he usually emerged successful. In 1789 the Inspection Report notes 'the peculiar circumstances' that the regiment has a fixed recruiting quarter in Yorkshire—a happy association which has lasted to this day, although the main recruiting centres are now in the South and West—which led to its being pretty sure of being well recruited.

By 1790 all seemed well and the Inspecting General observed:

> 'This is a highly finished Regiment. The men are graceful and excellent, with a martial countenance and manner that exhibit an exalted state of discipline';

and in 1792 George III expressed high approval of them.

Far more important to the ordinary soldier, however, than any bouquets about their appearance was the grotesquely overdue introduction in 1792 of the first army barracks—though the cavalry had to wait a few more years for its turn. As for the comparatively few wives of 11th Dragoons, they would have to wait over half a century for the privilege of married quarters, and even then, in 1848, such were only barrack rooms. At the end of the eighteenth century it was still normal practice for a married man to annex a corner of the barrack room, which became known as 'the marriage corner'. It was part of the wife's duties to wash, mend and sometimes cook for the whole barrack room; the married corner system was regularly recognised by authority, the number of women in barracks being six per troop or company, reckoning one wife to eight dragoons or twelve infantry!

Chapter 3

'In England I never saw nor heard of cavalry taught to charge, disperse and form, which, if I only taught a regiment one thing, I think it should be that. To attempt giving men or officers any idea in England of outpost duty was considered absurd, and when they came abroad, they had all this to learn. The fact was, there was no one to teach them. Sir Stapleton Cotton tried, at Woodbridge in Suffolk, with the 14th and 16th Light Dragoons, and got the enemy's vedettes and his own looking the same way.'

(Lt-Col. William Tomkinson, *The Diary of a Cavalry Officer in the Peninsular and Waterloo Campaigns.*)

AFTER thirty years of uneventful home service, the regiment was now due for action; furthermore, it was practically a new regiment since its conversion into Light Dragoons. Dundas had not been wasting his time these years of peace, and not only a good spirit of discipline had been established but the beginnings of the true regimental spirit in which the close and confident relationship between officers and men is the predominant factor. It was true that pay and pensions were still wretched, that for the common soldiers hardships abounded and rewards seemed few, that discipline by today's overly sophisticated standards was rigid to the point of ferocity, but already there was comradeship that could sometimes hurdle barriers of rank, pride of place and reputation, and a common solidarity against the harsh and miserly dictates of Westminster and Whitehall. But the status of the soldier was still false and absurdly low at the end of the eighteenth

Army Museums Ogilby Trust
A water-colour, dated 1796, showing a 'British 11th Light Dragoon'.

century—little better than it had been in the days of Marlborough. Fortescue puts the position:

> 'Marlborough's troops marched from victory to victory, and returned to be cursed as the plagues of the nation. Thousands of officers and tens of thousands of men were turned adrift. The only result was to make the survivors cling more closely together. Maltreated and despised sects – and such were the regiments of the eighteenth century – are likely, especially if they be English, to gain vitality rather than to lose it.'

Fortunately this was so in the Eleventh; and fortunately too, because it was growing into a smart cavalry regiment, many of its officers did not lack money or hesitate to use it for the good of their troops.

> 'In regiment after regiment,' Fortescue adds, 'officers had thought out the means of bettering the soldiers' lot, and thereby were making good the meanness of the State.'

In France, great and terrible deeds of infinite consequence were now lighting up the stage of history, with the storming of the Bastille in 1789, the adoption of the tricolour, peasant risings in the country and mob rule in Paris, and all the fiery hope, bloodshed and intrigue of the French Revolution. Austria and Prussia, encouraged by France's internal strife, allied themselves against her in 1792, with the result that the Girondists declared war against them, and on January 21, 1793, exactly four months after the monarchy had been abolished and France had become a Republic, Louis XVI was executed. Eleven days later the National Convention government declared war against Britain, Holland and Spain, and annexed Belgium. This French occupation was halted when the Austrians recaptured Brussels in March, and at the end of April two squadrons of the 11th Light Dragoons, under Major George Michell, joined the Duke of York's Army in Flanders, fighting alongside Prussians and Austrians. The regiment now disposed of nine troops, a second Lieut-Colonel and a second Major having been added to its establishment.

On May 24 the French crossed the Escaut and threatened Denain. Captain Charles Crauford, A.D.C. to the Duke of York, spotting an enemy baggage-train near the river, galloped off to Colonel Dundas and suggested that the Eleventh make a surprise attack on it. This, despite superior French numbers and the risk of being cut off, they executed with dash, Crauford at their head, and Cannon records:

> 'Overtaking the enemy's baggage-guard, the Eleventh rushed upon them sword in hand with terrific violence, broke them in an instant, sabred fifty men on the spot, took fifty-six prisoners, and captured eight waggons laden with baggage, and thirty horses. A strong body of the enemy advanced to cut off these daring British troopers; but the Eleventh effected their retreat with the loss of three men and three horses.'

The Duke of York, having expressed warm approval of this small action (his Adjutant-General's despatch refers to twenty-eight waggons captured), turned his attention to the siege of Valenciennes, which lasted forty-five days, and then to that of Dunkirk, which had to be raised; the regiment was present at both. But it was on April 26, 1794, in the area round Le Cateau that the British cavalry at last came into its own. The previous day two squadrons of the 15th Light Dragoons and two of the Austrian Leopold Regiment under General Otto had not only routed the French cavalry near Villers-en-Cauchies but had broken an infantry square, inflicting some 1,200 casualties to only fifty-one of their own. Now General Chappuis, determined on revenge, advanced on Le Cateau in two columns totalling 25,000 men and seventy-nine guns, under cover of a dense fog. They soon drove back the British outposts in the villages in front of Le Cateau heights, from which the Duke of York's main body could hear distant pistol shots. But heavy and accurate British artillery fire caught the advancing French off guard, and their two columns were forced to pull back, becoming entangled in the process. Now the sun broke through, and the Duke of York exploited the enemy's confusion by ordering his entire cavalry into the attack, so as to turn the enemy flank while smashing his front at Bethencourt with infantry and artillery. General Otto sped off with the Austrian Cuirassiers and two British cavalry brigades, and ran into 2,000 French horse and infantry, with fourteen guns, near Caudry. Many of these fled, but the infantry guns were captured, and Otto continued his pursuit until his force came upon the rear of Chappuis's division near Audancourt, where it took many prisoners and fourteen more guns. Then the Brigades of Mansell and Vyse attacked Chappuis himself near Montigny, capturing him and all his guns. The cavalry engagements followed each other in quick succession, so fast that the British infantry were unable to join in; when a large enemy column was scouted moving in the Prêmont-Marets area Major Stepheicz, the Austrian officer commanding a joint force of his own Hussars and the 7th and 11th Light Dragoons, ordered an immediate charge and followed up its initial success to gain a decisive victory over greatly superior numbers. Cannon refers to 'the most heroic gallantry' of this Austro-British force, recording that:

> 'The brave troopers used their broad swords with such energy and effect, that twelve hundred Frenchmen lay on the field; and ten pieces of cannon, with eleven tumbrils filled with ammunition, were the trophies of this display of British valour. The impetuosity of the charge was so irresistible, that the French were instantly broken, and the only loss sustained by the Eleventh was five horses killed and two wounded. The Duke of York complimented the cavalry on their distinguished conduct, and particularly mentioned the determined gallantry with which the Seventh and Eleventh Light Dragoons charged the superior numbers of the enemy on the left.'

During the whole day's cavalry actions (in which Sir Stapleton Cotton was blooded) the French lost at least 4,000 men and thirty-five guns—the English two officers and fifty-four men killed and four officers and 100 men wounded. Fortescue, in his *History of the British Army*, puts the French casualties even higher, and takes pains to emphasise the pinnacle of performance reached, in his opinion, in this now obscure battle in north-eastern France, ironically today quite forgotten.

> 'So ended the greatest day in the annals of the British horse, perhaps the greater since the glory of it was shared with the most renowned cavalry in Europe. The loss of the Austrians was nine officers, two hundred and eighty-nine horses, killed, wounded, and missing. The

British regiments that suffered most heavily were the Blues and the Third Dragoon Guards, each of which had sixteen men and twenty-five horses killed outright. . . . The total loss of the covering army was just under fifteen hundred men; that of the French was reckoned, probably with less exaggeration than usual, at seven thousand, while the guns taken from them numbered forty-one.'

Chappuis had been captured with all his private papers and dispositions, in the light of which the Prince of Coburg and the Duke of Kent reinforced St Amand with thirteen battalions of foot and six or seven regiments of cavalry. After the fall of Landrecies the Duke of York took up a covering position near Tournai, where the French with 30,000 troops under the Dutch General, Daendels, attacked on May 10; Lieut-General Harcourt, with sixteen squadrons of British and two of Austrian cavalry managed to turn the French right flank, which withstood six deep their first charge. The second, however, broke the French line and the allied cavalry demoralised their enemy, killing 4,000 Frenchmen, and capturing 400 men and thirteen guns. The Eleventh lost only seven men killed and three wounded, nine horses killed and eight wounded.

Later in May the Austrians suffered several reverses, notably at Fleurus, as a result of which the British had to evacuate Tournai, and finally, though French attacks on Alost and Malines were unsuccessful, the allies were forced to abandon the Austrian Netherlands and retire across the Meuse into Holland. But of that great day of May 10 Fortescue wrote:

'Not for eighteen years was the British cavalry destined to ride over French battalions as they rode on this day; and then Stapleton Cotton was fated once more to be present, leading not a squadron of Carbineers, but a whole division of horse to the charge at Salamanca.'

In January 1795 the regiment was part of a British army defending the line of the river Waal against vastly superior French forces; this was feasible until severe frost allowed the French to cross the ice near Bommel and turned the whole country into a bleak white plain. There were delaying actions at Geldermalsen and Meteren, but the enemy's far greater numbers and their welcome by the Dutch, who would do nothing to help the British, resulted in a precarious and painful retreat through Holland into Germany, and the surrender of Flanders to France. The troops recovered in Germany, and before their return to England the regiment, with the rest of the light cavalry, spent some time both in the Westphalian plain and at Stade, northwest of Buxtehude and Hamburg—both areas to become intimately known to it exactly 150 years later. It embarked at Tielenfleth on the Elbe on December 16, but owing to appalling weather and consequently an enforced stay in Glückstadt, a possession of the King of Denmark twenty-eight miles from Hamburg, it did not reach Gravesend until February 15, 1796.

The regiment, now commanded by Lieut-Colonel J. W. Childers, was back in Holland in September 1799, again under the Duke of York; Prussia, at odds with Austria, had withdrawn from the war in 1795, but now Russia had come in, and there was an ambitious threefold plan: an Anglo-Russian army under the Duke of York would drive the French from the Netherlands (although most of the Dutch showed no wish to be 'liberated'); an Austrian army under Archduke Charles would expel the French from Germany and Switzerland; and a Russo-Austrian army do likewise from Italy. This tall order never looked like being effected, the more so since the young artillery officer Napoleon Bonaparte, first distinguished in 1793, was now fast approaching the peak of his brilliance and power as a great commander.

At all events, the regiment landed in September after its

first taste of amphibious operations—for most of the horses had to be swum ashore on to the island of Texel in northern Holland. The campaign that followed, despite isolated successes, was an utter failure, mainly because there was a breakdown of understanding between the Russians and the English (the ablest Russian general, D'Hermann, was captured by the French), and because the bulk of the Dutch people if not openly hostile were apathetic. Indeed the Dutch troops under Daendels and the French under Brune more than held their own.

This marked the first and only time that the regiment had fought alongside Russian troops. It was usually split amongst two brigades, and on occasion a squadron might actually come under Russian command. In one October attack in the sandhills between Bergen and Egmont-op-zee directed by Lieut-General David Dundas an Eleventh squadron turned the flank of an enemy position, while the 55th Foot and a Russian light battalion stormed the breastworks from the front; in the process the squadron nearly had to swim its horses fording deep water. A few days later two squadrons were attached to the Russian infantry attacking Wijk-op-zee and the other two were directed on Beverwijk. Again the losses were heavy, exceeding 2,000 killed or wounded on both sides, and the regiment lost ten men killed and several wounded. On October 17 hostilities were suspended, and five days later the Russians withdrew from the coalition in disgust, being particularly annoyed with the Austrians. Under the convention of Alkmaar (round which there had been heavy fighting) the Anglo-Russian army had to evacuate Holland by the end of November, and all prisoners on both sides were exchanged. It seems typical of a thoroughly botched campaign, redeemed only by the bravery and dash of the British participants at low level, that perhaps for the first, but certainly not for the last time went up the anguished cry, 'What, no transport?' when the regiment was supposed to embark from Den Helder. Ship space being short, Colonel Childers, who had already lost sixty-two horses in battle or from sickness, had to order his men to shoot another 152 on the beaches—the very worst decree for a cavalryman. It is an illuminating pointer to the esteem in which the Russians were held that only fifteen chargers were given to them.

* * * * * *

The new century started auspiciously when General Sir Ralph Abercromby, who had seen something of the calibre of the regiment in the late campaign in Holland, honoured it by asking especially for a detachment to serve in his expedition designed to drive Napoleon's troops from Egypt in 1800; his original cavalry consisted of the 12th and 26th Light Dragoons and Hompesch's Hussars, but four officers and seventy-five men from the Eleventh, under Capt.-Lieutenant Money, were thus added to the strength. Although the expedition sailed in the summer, it was held up by so many delays, misfortunes and storms that it was not until March 8 that it managed to land at Aboukir Bay, under heavy fire of grape and musketry at that. Abercromby's troops dislodged the French and advanced several miles, but the non-arrival of most of the horses for some days prevented effective reconnaissance and the exploitation of an otherwise successful action that Abercromby fought on March 13, 1801, with severe casualties on both sides. The expeditionary force was fortunate to have with it Major-General John Moore, the light infantryman who had done so much to introduce reforms and new thinking into the British Army, partly in the light of lessons learned in the disastrous American campaigns. Not all these had penetrated yet to the cavalry, who still tended to impatience and disregard of efficient outpost and scouting techniques. At all events, in their first engagement on March 18, a day after the surrender of the castle of Aboukir, the two British regiments under Colonel Archdall failed to detect an ambush of French grenadiers near Beda on the Alexandria Canal, in their enthusiasm for charging the French cavalry, and were severely mauled. They were not helped by the fact that many of their newly arrived Turkish horses were only half-broken.

On March 21 General Menou attacked, and the battle of Alexandria took place, resulting in a decisive defeat for the French, who lost 1,700 men including three generals; the British had sixty officers and 233 men killed, sixteen officers and 1,190 men wounded, and at one time the fighting was so fierce that regiments on both sides ran out of ammunition, soldiers of the 28th and 42nd Foot distinguishing themselves with bayonets and even stones against French cavalry sabres. Abercromby, who had conducted himself with gallantry and distinction, was severely wounded and died a week later, the command passing to General Hutchinson.

Wavering Turks and Bedouin, for what they were worth, threw in their lot now with the British, but the force was still short of camel and horse transport and of water; it had found out, as Major-General von Ravenstein was to describe it so aptly 140 years later, that while the desert might be a tactician's dream, it was a quartermaster's nightmare. By May 9 command of the Nile waterway had been secured with the evacuation of Rahmanieh by its large French garrison, and now General Menou's Alexandria-based army was split from that of General Belliard at Cairo. All eyes in the expeditionary force were now set on that city. Cairo, as usual, was riddled with intrigue, and most of the Mamelukes were only too ready to desert the French; but before preparing his plan of attack General Hutchinson sent two officers of the Eleventh, Capt.-Lieutenant Money and Lieutenant Lutyens, on a patrol to spy out the city's defences, which they creditably did at close range despite French pursuit. But with native desertions and the arrival of more British and Indian troops from Bombay a long siege proved unnecessary and General Belliard capitulated in June; on August 30 because of lack of provisions Menou followed suit at Alexandria, last remaining outpost of French power in Egypt.

This first desert campaign had taught some valuable lessons to C Squadron, 11th Light Dragoons, from which came the eighty members of the regiment who took part in it. But it had not been suitable country for poor mounts, and the major difficulty had been the impossibility of providing the cavalry with adequate forage.

Since 1802, when C Squadron returned to England, the regiment has borne on its guidons and appointments the Sphinx's emblem and the word 'Egypt'—and C Squadron has always been in the position of honour, right of the line, on ceremonial occasions. In more modern days it has even been rumoured that some married families instruct their children that the alphabet runs 'C, A, B, R, H, Q'!

Chapter 4

The Peninsula and Waterloo

'. . . the cavalry of other European armies have won victories for their generals, but mine have invariably got me into scrapes. It is true that they have always fought gallantly and bravely and have generally got themselves out of their difficulties by sheer pluck.'

(*Wellington*, quoted by Captain Gronow.)

ONE result of the French War had been a considerable increase in the strength of the regiment, but with the peace treaty of Amiens in 1802 it was reduced to slightly over 500 men. But in the summer of 1803, because of the real threat of an invasion from Boulogne, the regiment, now on the south coast, was enlarged again and by autumn was bigger than ever before, mustering ten Troops of ninety-five privates each. For war had broken out again in May, 1803. The 11th, however, were not to return to the Continent until 1811, their next foreign service being a first tour of duty in Ireland between 1807 and 1810, where they were stationed at Clonmel and Dublin, with periodic hunting down of highwaymen and illicit stills.

The Peninsular War, or the war against the French in Portugal and Spain, had already been in progress some three years when in June 1811 the 11th Light Dragoons joined Wellington's army between Elvas, in Portugal, and Badajoz, the Spanish town on the river Guadiana that Wellington had been besieging, so far without result. On their approach march from Lisbon the regiment, now 725 strong all ranks, had met a convoy of some 700 wounded survivors of the battle of Albuera, which they had narrowly missed. They had also, of course, not been present at the earlier battles of Vimeiro (1808), Corunna and Talavera (1809), or at the lines of Torres Vedras in the winter of 1810–11. Wellington had still not been able to take Badajoz, and when two strong French armies, one under Marshal Soult, victor of Corunna, the other, known as the Army of Portugal, under Marshal Marmont, advanced to raise the siege, Wellington gave it up and took up new positions on the smaller river Caya, with his left bounded by Campo Maior, his right by the Guadiana. By June 21, 1811, the regiment, as part of Major-General Long's brigade with the 2nd Hussars, King's German Legion, was on the right of the army, bivouacked in the woods between Elvas and the Guadiana, with the winding Caya stream directly to its front.

As so often happens, when a regiment or a battalion joins a campaign late, either its enthusiasm to prove itself or its inexperience, or a combination of both, achieves not startling success but near disaster or ignominious loss. Thus it was with the Eleventh's first engagement, redeemed only by blind courage which at least caused the enemy substantial casualties.

The King's German Legion Hussars were posted along the Guadiana, but their picquets were surprised by the French cavalry and the Germans with considerable loss escaped to Elvas only with difficulty. According to Gleig, that fascinating Scot who himself fought in the Peninsula as a cavalry soldier and later became Chaplain-General to the Forces, and a prolific and greatly rewarding military historian, a German trooper had deserted to the French and given away all the Allied dispositions. That of the Eleventh, to say the least, had been unfortunate. The Caya was narrow and winding, with low and mostly open banks, and thus presented neither any serious obstacle nor a good viewpoint because of its many detours; on the British side, however, were some commanding hills not far away from the stream, and this is where Lieut-Colonel Henry Cumming should have placed his outposts. As it was, his picquets in squadron strength—120 men under the same Captain Lutyens who had distinguished himself in Egypt—were ranged along the low and angled ground of the river so that they could not possibly see an enemy until he was close at hand. Nor did they know the fate of their German allies, who had not warned them that the French were reconnoitring in force.

The French, according to Gleig, had lain up in a wood to the right of the British position during the night of June 22, and at dawn the next day attacked in superior numbers; this main body and the Eleventh fought for some three hours, but, unknown to Lutyens, another French force under Colonel L'Allemand had outflanked him and waited in his rear. Gleig, telling the story through the mouth of Private George Farmer, *The Light Dragoon* of his title, goes on:

> 'We never for a moment supposed these were Frenchmen; we took it for granted they were the Portuguese placed there for our support.'

Charged by the four French squadrons in front, however, the Eleventh were finally forced to give ground by sections, still skirmishing, and Lutyens ordered his squadron to retire upon what he took to be the Portuguese force. The mistake was not realised until only about a hundred yards separated the two bodies, and then Lutyens, *faute de mieux*, gave the order to close files and charge, a move which took the French second column, evidently expecting a surrender, entirely by surprise. Gleig continues:

> 'Never were men so utterly confounded. It was clear they expected nothing of the sort, as they stood still, looking us in the face, and never made a movement to

meet us. The consequence was, that coming upon them at speed, with all the weight and activity of our more powerful horses, we literally knocked them down like nine-pins. Over they went, the horse and rider rolling on the ground; we cutting and slashing as we broke through. But alas! for us, there was a second line behind the first, which behaved differently.'

Greatly outnumbered, the Eleventh were now cut off, and only those particularly determined *and* well-mounted, like Lieut. William Smith, could hope to escape, even when severely wounded as he was. The result of the day was near catastrophe: eight killed, twenty-two wounded, and two officers and seventy-five N.C.O.s and men taken prisoner.

According to Gleig's account of the incident the French lost even more than the British, since more than fifty wounded men were brought to the Spanish hospital where Private Farmer was treated. Gleig also relates among various gallant acts the extraordinary tale of Private Wilson, who, whilst engaging a French dragoon in hand to hand combat received a deadly thrust from a French officer which seemingly killed him instantly.

> 'Yet though he felt the sword in its progress he, with characteristic self-command, kept his eyes still on the enemy in front, and raising himself in his stirrups, let fall on the Frenchman's helmet such a blow that brass and skull parted before it, and the man's head was cloven asunder to the chin. Both he who gave it, and his opponent who received it, dropped dead together. The brass helmet was afterwards examined by order of a French officer, and the cut found to be as clean as if the sword had gone through a turnip, not so much as a dent being left on either side of it.'

Gleig notes that the French cavalry nine times out of ten used the point of their sabres, whereas the British struck with the edge. Napier also relates the episode, estimating the combined German and British casualties resulting from it at 150, adding 'the French aver that Colonel L'Allemand drew the British cavalry into an ambuscade'. Major Thomas Downman, of the R.H.A., recorded in his diary that the enemy had used 3,000 cavalry near Campo Maior, and noted that Captain Lutyens had apparently been charging some of their squadrons when he found himself surrounded by a very superior force. His entry ends: 'Lord Wellington is stated to be much vexed about it.'

Tomkinson, at that time a more experienced campaigner with the 16th Dragoons, of which he was a Captain, also recorded the affair, which obviously gained some notoriety in the army. He tells how the French had put a strong cavalry force across the Guadiana and got in the rear of the Eleventh, whose picquet was badly posted, and whose leader, Lutyens, had not been properly briefed on the ground (the old story of the poor hand-over of a position).

> 'It was probably the first picquet he ever mounted, and on seeing enemy's troops on the road he looked to for retreat, had not his wits quite about him.'

When all was said and done, however, it was an unfortunate start to the campaign for a regiment that only a few days earlier had joined others more experienced in the Spanish sun and dust. It was not surprising that Wellington, always on the lookout for rashness or indiscipline in his cavalry, was annoyed. As Tomkinson was to write some years later, English cavalry regiments knew little of field tactics and practised less at that time, merit being esteemed by 'celerity of movement', in other words dash. But dash without training, discipline and orderly movement was useless, as he realised, just as not a few battles in which Rupert's cavalry had swung the day and panicked the enemy had been drawn, not won, sometimes even lost, because he would not control his cavalry but let them rove, as it were, half way over Oxfordshire and Warwickshire.

Tomkinson, somewhat in advance of his time as a cavalry tactician (he advocated the establishment of a system of regular cavalry signals understood by all) wrote:

> 'We never teach our men to disperse and form again, which of all things, before an enemy, is the most essential. Inclining in line right and left is very useful, and this is scarcely ever practised. . . . But we go on with the old close column and change of position.'

But Gleig, who also went through it all, had perhaps the last word on the Caya disaster in *The Light Dragoon*:

> 'Then again, patrolling, which is an especial duty, puts the metal both of men and horses to the test. You must move forward as if you had a hundred eyes; you must be cool and collected, and prepared for every conceivable adventure. Neither hedges nor ditches must offer insuperable obstacles to your progress, whether you be required to take ground to the front or rear; and you must be quite as ready and as willing to gallop off when to convey intelligence in your business, as to fight with carbine or sword, where you are desired to delay an enemy's progress. In a word, both the light dragoon and his horse are called upon, as soon as they take their station in the front of an army, to acquire, as if by intuition, new ideas on every subject; for, except in the formation of column or line, and the act of breaking up into order of march, and closing into squadrons again, the home drill – at least in 1809 and 1810 – had not taught us much of our real duty.'

The British position in the Peninsula at this time was somewhat disquieting, in spite of previous victories; a second attempt at taking Badajoz had failed, the Spaniards and Portuguese seemed discontented allies, and Wellington's army was wasted by disease and hunger, for the raw Spanish countryside was not any easy one off which to live. The British were in fact now heavily outnumbered by the forces of Soult and Marmont, particularly in cavalry. It was rough country, and rough living, but good training for a regiment; when Wellington moved north against Ciudad Rodrigo the Eleventh, now part of Major-General Alten's brigade, became accustomed to doing outpost duty in front of the army. Even so, in the mountainous passes between the Estramadura and Old Castile, where Marmont had moved a substantial mixed force, another minor catastrophe occurred when the French surprised and captured a complete patrol of ten men under Lieut Frederick Wood at St Martin de Trebejo on August 15, 1811. It is more than possible that in this embarrassing incident lies the origin of the nickname 'Cherrypickers'—which dates traditionally from the surprise of a detachment of the Eleventh forced to fight in a cherry orchard some time during the Peninsular War.* Today no one seems to know exactly when, where and how the name originated, but it is none the less well known for that.

*Captain Gronow, in his reminiscences, is very forceful about this incident; he was writing a long time after it took place, and it is not clear how accurate his account is (though the main facts are not in dispute), since it appears in a passage mainly concerned with the retreat from Burgos, which did not take place until more than a year later. But what Gronow has to say is interesting:

'The system of outposts in the French Army was on a different footing from ours. Before the enemy, the French sentinel was relieved every hour; whereas our soldiers remained on duty two hours!—the extra hour caused great fatigue, and in cold weather induced sleep. A troop of the 11th Light Dragoons on duty in front—that is, at the extreme vedette, in the immediate presence of the enemy—were once caught napping. The French officer in command, observing the bad ground kept, ordered forward a sergeant and five men who entered our lines and found Captain Wood and his men fast asleep; when the dragoons awoke, they were compelled to surrender themselves prisoners of war. Now if the vedette had been changed ever hour, this disgraceful catastrophe would not have occurred. Doubtless all these matters are better arranged now: the Crimean war ought to have taught us many valuable lessons. . . .'

(*The Reminiscences and Recollections of Captain Gronow, 1810–1860.* Two volumes, John C. Nimmo, 1889.) Lieut Wood was severely wounded at Waterloo, and died in 1861.

By the latter half of September the two French armies of Marmont and General Dorsenne had linked up near Ciudad Rodrigo, having first been able to make contact at St Martin de Trebejo, their combined strength being some 60,000 men, of whom 6,000 were cavalry. Wellington could only call upon 40,000 (4,000 cavalry), of whom not many more than half were British, so he wisely withdrew from the flatter ground near the town to better positions on the higher ground round Fuente Guinaldo; thus two squadrons of the Eleventh, with one of German Hussars, in the 3rd Division took positions on the heights of El Bodon and Pastores, on the left of the Agueda River. The regiment's remaining two squadrons were posted at Gallegos and Ituera. The British now commanded a complete view of the disputed town of Ciudad Rodrigo, round which the enemy were milling on September 23 and 24 with large convoys of food, waggons and mules.

All through September there had been patrolling and skirmishing by day and night, as Captain Michael Childers described in a letter from El Bodon to his half-brother Colonel J. W. Childers. Six Troops of the Eleventh and 800 men of the 88th Foot were based at El Bodon, and another two Troops with Alten's brigade headquarters at Itrero. Captain Childers mentions to his brother the unhealthy state of Wellington's Army (19,000 sick on the last return, including more than 800 officers ill at Lisbon), notes that the French will pay three shillings a loaf for bread (an extraordinary price for those days), that most of the Spanish countryside has been burnt, the atmosphere of the country always being hazy from the fires (a substitute for manuring), and finally that the local shooting is excellent. The Colonel (Cumming), he and Lieut. Jenkins (who had joined the Eleventh by the then prevalent method of obtaining a cornetcy in a regiment by raising a certain number of men, probably from tenants), had shot some partridge and quail but missed ten hares.

Finally on September 25 Marmont sent fourteen squadrons of the Imperial Guards from Ciudad Rodrigo against the left wing of the allies, with indecisive results, and thirty squadrons of cavalry, fourteen battalions of infantry, and twelve guns against Fuente Guinaldo. Wellington therefore directed Major-General Colville to form the 5th and 77th Foot, the 21st Portuguese Regiment, and a brigade of Portuguese artillery, on the hill over which the Guinaldo road passed, their flanks being covered by the Eleventh and one squadron of German Hussars on the heights to the left of El Bodon.

The French cavalry, splendidly disregarding heavy fire from both the Portuguese artillery and the mixed infantry, swept across the dusty plain and charged up the hill, only to be charged and beaten back as they approached the crest by the British and German squadrons. This happened not once, but time after time, for the French were both courageous and greatly superior in numbers (ten to one according to Cannon, four to one, a more reasonable figure, according to Lord Londonderry of the 10th Hussars). Napier relates how the Eleventh and the German Hussars charged the French masses some twenty times, saying: 'It was astonishing to see so few troopers bearing up against that surging multitude.' Cannon refers to French resolution and to the 'astonishing efforts of valour' of those who effectively frustrated it, as charge succeeded charge 'for upwards of an hour'. He particularly commends the handling of their squadrons by Captains Childers and Ridout respectively, and if all behaved gallantly Childers and T.Q.M.S. Hall stood out for their daring courage. Hall, who three years later was promoted Quarter-Master with officer status, was wounded, and Childers, as he told his half-brother in an elated note after the battle, had his mare twice cut by sabre thrusts and once hit by a spent bullet. All his officers were either wounded themselves or had their chargers killed or wounded. The French did succeed in capturing some

Portuguese guns early on, but the 5th and 77th Foot, on being charged in turn, 'drove them down the hill like a flock of sheep and retook the guns'. Childers describes how his own squadron took part in eight charges over an hour-long period, 'each time 5 or 6 times our number'.

Their position on the high ground naturally favoured the allied cavalry—provided they did not charge too far—but of course they had to reform and regain the high ground after each charge, and so strong in numbers were the French that it was an exhausting process. Childers notes that when his squadron drove the French back on their main body 'we of course did not follow them down the hill but retired about 100 yards', and then repeated the process; according to Cannon, however, one squadron of the Eleventh did charge too far and got entangled in the difficult ground of the ravine across which the French had originally passed to attack the heights. But it evidently extricated itself. When Childers's squadron made its last charge it was down to about twenty men, who got so mixed up with the French, who managed this time to stand firm, that after a mêlée the British had to retreat at the gallop for some three quarters of a mile. The hotly pursuing French were then stopped short by a solid square of British infantry from the 5th and 77th Foot, firing at extremely close range, and the horsemen by this time were so indiscriminately mixed up that this fire killed more than one British and German soldier. Childers reported in his third letter:

> 'The conduct of the Infantry was admirable. Exactly the same steadiness as in a Barrack Yard, our men really behaved very well, and General Stewart the Adjutant General who was with us nearly the whole time and certainly behaved nobly, came back and as Cummings was not then in the way told me he was sent by Lord Wellington to thank us. As many Columns of Infantry were halted on the hills and Lord Wellington within a

> quarter of a mile, it was all seen, and really we were given more praise than you can imagine (all this *entre nous*). We then retreated with the Guns and Solid Columns of Infantry to Fuenti Guinaldo, about 5 miles, the most correct field-day could not be more beautiful, we had the good luck to save the baggage of the 88th, 74th, 45th which the French had taken, but we sent down about 20 men and got it back. The French followed us with their Cavalry and Artillery but kept at a respectful distance. . . .'

In all the Eleventh lost ten killed that day and twenty-one wounded, including their colonel, with nine horses killed, twenty-six wounded. Horses at this time were commonly suffering from what Childers describes as 'fever in the feet', but soldiers were as likely to go down with disease, particularly 'the ague', as to become battle casualties. Thus at El Bodon two-thirds of the officers of the regiment were unable from sickness to fight.

> Tompkinson in his account of the battle says: 'The 11th were particularly steady, and the whole got off with very little loss. Lord Wellington was there, and aware of the critical situation they were in. Our infantry behaved in their usual cool way, and the conduct of the 21st Portuguese was much spoken of. Captain Childers, of the 11th, was the officer who particularly exerted himself, to which the safety of the detachment is to be considerably ascribed. He charged bodies three times their number, rode at them with the greatest determination, and always succeeded. The conduct of the 11th Light Dragoons was such as must stamp them as soldiers doing their duty in a critical situation. Very flattering orders to the troops came out.'

Wellington, certainly not a general given to loose or even habitual praise of his troops, in fact issued a general order holding up the conduct of the soldiers of Colville and Alten as an example to the whole army, especially because of the great odds against them. It must have been a welcome touch indeed that in this public commendation the behaviour of Lieut-Colonel Cumming was particularly noticed. El Bodon had wiped out in full the shame of two unfortunate patrol reverses.

Wellington's men spent an uneasy night, and fully expected another big French attack the next morning; but despite a superiority of four to one and the threatening and incessant beating of drums it was evident that the French did not appreciate Wellington's weakness, and Marmont delayed his attack. This was fatal, for the next night Wellington slipped away to fight another day, his force withdrawing in such good order that not a straggler or a piece of baggage was left behind.

But by the end of September Wellington was again investing Ciudad Rodrigo, which eventually fell on January 19, 1812. The Eleventh spent the winter first in the Spanish Estramadura, then on the Tagus, and finally in the Portuguese Estramadura, but in March returned to patrol action in the area round Badajoz, which was besieged and fell on April 6. All British cavalry regiments in Spain had now been reduced to three squadrons. At this time also the regimental uniform was altered to a coatee with buff facings and silver epaulettes; pantaloons, short boots and cloth overalls; hussar saddle and blanket, blue horse furniture with a silver lace border; and instead of a helmet, a shako.

On June 11, 1812, Wellington's army, now reinforced to a strength of 24,000, crossed the Agueda and advanced towards the Tormes, across which river Marmont withdrew on June 16 when the British reached Salamanca and besieged the forts, which fell by June 27, when the French withdrew again towards the Douro. By July the opposing armies were on either side of that river, the 11th now being in Major-General George Anson's brigade with the 12th and 16th Dragoons. On June 30 the Eleventh had lost two sergeants and a private taken prisoner near Alaejos, when they were still in Alten's brigade. According to Tomkinson Sir Stapleton Cotton, who commanded all the cavalry, was 'much vexed at the manner they were lost, and said the 11th should leave that brigade'. Tomkinson adds 'I do not know that he could say so with reason.' Marmont was reinforced by General Bonnet with 8,000 men and crossed the Douro on July 16, forcing Wellington's cavalry to withdraw first to the Trabancos river, and then to the Guarena, upon the town of Canizal. The Eleventh distinguished themselves in various skirmishes.

For the rest of the summer and autumn of 1812 the Eleventh were consistently active and often in action; they fought various successful engagements after taking part in the great victory of Wellington over Marmont at Salamanca on July 22, leading the pursuit to the Douro, and entering Valladolid on July 31, where the French had abandoned a vast quantity of stores. Amongst these successes in small half-forgotten villages and towns were the defence of a bridge and ford at Tudela on the Douro; the capture intact of a French patrol at Cisteringa near Valladolid, which General Picton described as 'the quickest thing he ever saw cavalry do'; an affray against odds at Torquemada; the timely relief of the 12th Light Dragoons under extreme pressure at Monestario and the saving of that town, not far from Burgos—which had come under siege on September 20; and a fine rearguard action at Cellada on the Pisuerga river after the siege of Burgos had been raised a month later and the British had to retreat to the Douro again. The retreat from Burgos was a sad business, though not entirely disastrous, but although the Eleventh had lost nearly fifty all ranks killed, wounded and missing, and forty horses, on the Pisuerga, in the final stages of the retreat between Salamanca and Ciudad Rodrigo they lost only a few men and horses. Further, they had again been praised by Wellington, and, this time, by Cotton.

One curious action, however, does deserve more detailed description. This took place at Castrejon on the Trabancos river on July 18, nearly a month after Salamanca, and was notable for the confusion that developed and the resulting danger to high-ranking officers including Wellington himself.

Cotton had placed his cavalry in rather exposed positions, but had maintained them all day with skill and resolution, so Napier tells us. At seven in the evening Wellington, accompanied by General Beresford, arrived to look at the enemy's dispositions. 'At this moment,' Napier goes on,

> 'some French horsemen, not many, broke suddenly away from the head-land beyond the Trabancos and came galloping on as if deserting, but soon with a headlong course they mounted the table-land on which Cotton's left was posted, and drove a whole line of cavalry skirmishers back in confusion.' [These belonged to the 12th Dragoons.] 'The English reserves on that side then advanced from Alaejos, and these furious swordsmen, scattered by their own charge, were in turn driven back or cut down; yet thirty or forty led by a noble officer, brought up their right shoulders and came over the edge of the table-land, above the hollow which separated the British wings, at the instant when Wellington and Beresford arrived on the same slope. Infantry picquets were in the bottom, and higher up near the French were two guns covered by a squadron of light cavalry disposed in perfect order. When the French officer first saw this squadron he reined in his horse with difficulty, his troopers gathered in a confused body round him seemingly as lost men, and the British instantly charged; but with a shout the gallant Frenchmen soused down upon the squadron and the latter turning galloped through the guns; then the whole mass, friends and enemies, went like a whirlwind to the bottom, carrying away Lord Wellington and Beresford, who with drawn swords and some difficulty got clear of the tumult. The French horsemen, when quite exhausted,

were attacked by a reserve squadron [of the 11th] and most of them killed, but their indomitable leader, when assailed by three enemies, at once struck one dead from his horse, and with surprising exertions saved himself from the other two, though they rode hewing at him for a quarter of a mile.'

Sir Charles Oman tells much the same story:

'Wellington was involved in person in the end of the cavalry bickering, and in no very pleasant fashion. He and Beresford, with their staffs, had arrived on the field about seven o'clock, in advance of the two heavy cavalry brigades, who were coming up to reinforce Cotton. He rode forward to the left of the skirmishing line, where two squadrons, one of the 11th and one of the 12th Light Dragoons, were supporting two guns of Ross's troop [R.H.A.] on high ground above the ravine of the Trabancos river. Just as the Commander-in-Chief came on the scene, a squadron of French cavalry, striking in from the flank, rode at the guns, not apparently seeing the supporting troops. They met and broke the squadron of the 12th Light Dragoons, which came up the hill to intercept them. Some of Beresford's staff, seeing this, and conceiving the guns to be in danger, rode up to the retiring squadron calling "Threes About!". This unfortunately was heard by the supporting squadron of the 11th, who, imagining the order to be directed to themselves, went about and retired instead of advancing to relieve their broken comrades above. Therefore the mass of pursuers and pursued from the combat on the flank, came hurtling down on the guns, and on the headquarters staff just behind them. Wellington and Beresford and their followers were swept away in the rout, and had to draw their swords to defend themselves. Fortunately the misdirected squadron of the 11th soon saw their mistake; they halted and turned, and falling on the scattered and exhausted French dragoons drove them back with great loss; few, it is said, except the *chef d'escadron*, who showed uncommon gallantry, got away. It was a dangerous moment for the allied army - a chance thrust in the *mêlée* might have killed or disabled Wellington, and have thrown the command into the hands of Beresford or Stapleton Cotton.'

In the twentieth century, perhaps to the detriment of the entertainment of the troops, General officers have been necessarily less prone than their forbears to get caught up in direct military action. Tompkinson adds to the story the fact that the French had wheeled up sixteen guns against the British outposts, which had obliged the R.H.A. troop of Major Ross to retire by accurate shooting.

> 'It was the sharpest cannonade, for the time, we were, or I was ever exposed to, and almost impossible to get the men away in complete order. Many shots went over us and struck the 11th Light Dragoons in the rear.'

Tomkinson confirms the confusion that arose from the 'Threes About!' order, but sums up 'There was no harm done, and our dragoons [the 11th] immediately advanced and drove them back.'

At the end of October 1812 Wellington's whole army went into winter quarters in northern Portugal; in the spring of 1813 it moved back into Spain, but to its great disappointment the Eleventh was not to participate in the final victorious campaign that was to drive King Joseph and the French back across the Pyrenees into France itself after the battle of Vittoria. For in March the regiment was ordered back to Portugal and thence home because of its losses both in battle and from sickness, and distributed its horses amongst other British and German cavalry regiments. In rather under two severe years its casualties had been 417 all ranks and 555 horses. Tomkinson, the 16th Dragoon, noted the date of April 3, 1813 as the day the Eleventh gave up their horses near Coimbra and left for England, and added his own summing-up: 'I think there are other regiments not so efficient as the 11th, but the man at their head is an old officer, and in the way of others more deserving to command brigades.'

Colonel Cumming (he had been promoted full Colonel in 1812) was in fact only 42, even if he appeared more venerable, and had drawn compliments from Wellington. But he appears to have been a difficult man. It so happened that for a time there were two Lieut-Colonels in the regiment, the junior and more popular being James Wallace Sleigh, and this never makes for an easy situation. According to one of Michael Childers' letters Cumming mistakenly thought that Sleigh was working against him, and therefore behaved 'in a most extraordinary way' and looked to be riding for a fall. Nothing apparently came of it, and when Cumming for a time went sick and had to return to England Sleigh commanded the regiment ably until his return. Both men later became full Generals.

Nearly two years of hard and rough campaigning in Spain had taught the 11th much. Their record had not been without its blemishes, some considerable, but on the whole it had been a good one; and much had been learned, many improvements made, often the hard way through bitter experience. If the beginnings of professionalism went back to the guiding hands of able and veteran officers first of Marlborough and later of Cumberland, it took the Peninsular campaign to turn a largely raw, incorrectly or inadequately trained cavalry regiment like the 11th, however dashing and smart, into the highly regarded professional body that it had become by the time it disembarked at Portsmouth on June 17, 1813. Major-General Anson, the Cavalry Brigade commander, had written in March to Colonel Cumming of his 'extreme regret and concern' at the news of the departure of the 11th for England, and had referred to the 'exemplary behaviour and meritorious conduct' of 'this excellent regiment' under his command. Stapleton Cotton, commanding all Wellington's cavalry, forgot one or two black marks he had previously noted and wrote, 'I regret very much losing the remains of a Regiment which has, at all times, conducted itself so much to my satisfaction, and which I should be particularly gratified by having again under my command.' (This wish was to be granted in the Waterloo campaign.) But perhaps the greatest compliment had been paid in April, 1812, when Welllington, answering a request from General Hill for another Brigade to garrison Salamanca, had replied that Hill could have any troops except Anson's Brigade or the 11th, 12th and 16th Light Dragoons, whom he wished particularly to retain with himself.

It is undeniable that in many cases the British troops, and especially their young officers who had not seen active service, were outclassed by the formidably experienced, better trained and led French veterans of Napoleon—officers and men who had marched over and conquered half of Europe. Gronow was angry at first when he heard General Picton—himself risen from the ranks—say that French officers were superior to British ones in the Peninsula, though the British N.C.O.s and men were the finest in the world, but he later came to agree. The French General Foy thought so too; Napoleon and his commanders were more ready to promote men on pure merit without the same inhibitions of privilege, rank and money, though nepotism existed in the French as in any other army. Furthermore their training methods and tactics were more efficient. Major the Honourable C. S. Cocks, 16th Light Dragoons, Wellington's favourite outpost officer, wrote: 'The service of Light Cavalry, in which I have been always bred, is nowhere taught in England, and beyond a few practical rules can only be learnt on service by reflection and observation in the field.' Far in advance of his time, Cocks considered that the military education for young men should include at all costs modern languages (French indispensable), geography, economics, government, military history (the Seven Years' War and the 'late war, from the year 1792'), with geometry and the study of fortification highly desirable.

Cocks wrote to his Uncle from Spain:

> 'Perhaps you will be surprised at what I am going to say, but I most highly recommend a certain knowledge of music, dancing and drawing, or at least some two of these accomplishments, and fencing, too, is a capital exercise. When first a soldier becomes a prey to ennui it is all over with him. He is first sorrowful and then sick. But a man will always get ennui unless he has the power of amusing himself, for we cannot think for ever, especially at nineteen.' He added pointedly, 'It is worse when a young man is left by himself without much duty as sometimes happens, especially with Light Cavalry.'

Apart from a more professional approach to—and more experience in—cavalry tactics, outposts, and manoeuvre at the beginning of the nineteenth century, the French held an advantage because their infantry were more used to and more adept at forced marches. It was not until men like Moore and Picton had learned and taught hard lessons that the British, whose N.C.O.s and men were potentially the finest in the world, redressed the balance. Gronow relates how

> 'The French soldier marched quicker than the English, both in advance and retreat; and after a victory by our troops few prisoners were taken. The Duke of Wellington, with all his wonderful foresight and genius, could never get at the secret why so few stragglers were met with in following the enemy; whereas at Burgos, after our raising the siege of that town, indescribable confusion arose, and nearly half the English army were either left behind or taken prisoners by Soult and Clauzel.'

If there were still too many gilded amateurs leading the British Army, for the cavalry private in and out of the line discipline was tough, sometimes unbearably harsh. Yet it was taken for granted in those days, and in the cavalry more than in most arms relationships in a good Troop, often then as now operating on its own for days at a time, could become close and warm between all ranks. Regimental Orders surviving from those days contain some pointers to what life in Spain and Portugal was like. Thus on April 11, 1812: 'Any man whose horse's back is hurt on the march in future, will be ordered to walk and carry his kit.' At Guinaldo on May 3 'The Out Troops are hereby informed that Robert Bull, Private in Captain Latour's Troop, has been tried by a Regimental Court-Martial for selling his horse's corn. He was found guilty and sentenced to receive 500 lashes. 205 have been inflicted only, the Surgeon being of opinion he could not receive any more.' At La Seca on July 6: 'Great inconvenience having been experienced from the few women that came out with the Regiment (and for the purpose of washing) being now employed in cooking for Officers, such Officers are requested to make arrangements as to enable these women to perform the business for which they were allowed to come out with the Regiment.'

Sir Stapleton Cotton more than once expressed concern about the bad state of arms, especially swords, in the cavalry, since, despite repeated orders to the contrary, privates used their sabres to cut wood for fuel. Food was a continuing problem in Spain, and forage for the horses; the wretched Spanish and Portuguese peasants, who were after all chiefly concerned to rid themselves of their French oppressors, tended, like most illiterate, poor and untrained groups to be maligned and mistreated by both sides, their crops and scanty livestock appropriated at will and their farms and poor hovels alike looted and burned. On their own side dishonesty and treachery were admittedly met too often. A Regimental Order dated August 1, 1812, at Culeas states: 'The followers of the Army, the Portuguese Women in particular, must be prevented by the Provost from plundering the Gardens and Fields of Vegetables. The Women must be informed that they must obey orders, or they will be turned out of the Army.' As always, the presence of *vitandières* brought problems as well as solace. Serious too was the menace to young unhardened soldiers of unfamiliar wine and raw spirits, which in hot weather caused much sickness and led to orders that they must be mixed with four parts of water. On August 20, for example, at Tudela, two men of the regiment were punished for drunkenness and the death penalty was threatened for future offences.

In the summer of 1813, after a brief spell in Ireland, the regiment was brought up to a strength of 759 horses and men at Hounslow, and discontinued wearing buff breeches. When Pte Farmer rejoined it about the end of 1813, Gleig has him say: 'I do not think that out of the five hundred men, from whom the fortune of war had separated me three years previously, one hundred continued to wear the uniform of the 11th Light Dragoons.' In 1814 the successive triumphs of Wellington and Britain's allies culminating in the first capture of Paris, forced Napoleon to abdicate the French throne, and the Bourbon Louis XVIII returned from his English exile to become—briefly—King. The 11th escorted Louis when he made a State entry into London with the Prince Regent on April 20, and provided escorts also that year for the Emperor of Russia and the King of Prussia. When at last Colonel Cumming, who had been present at every engagement in the Peninsula except the siege of Badajoz, was promoted Major-General, Lieut-Colonel Sleigh took over command. The right to bear the word 'Peninsula' on guidons and appointments was granted by royal decree.

The brilliant, lavish and disputatious Congress of Vienna was interrupted by Napoleon's unexpected return to France from Elba on March 1, 1815, followed within a fortnight by the flight of Louis XVIII to Flanders, and then by Napoleon's triumphal entry into Paris and the start of The Hundred Days on March 20, 1815. England's reaction was immediate, and war with France was again declared; six Troops of the 11th, amounting to 453 all ranks and 396 horses, disembarked at Ostend on April 2 and 3, the horses having to be swum half a mile ashore, and on April 6 the regiment witnessed the entry into Ghent of Louis XVIII—for the second time a fugitive from Paris. At Eyne, near Oudenarde, on the Scheldt (names familiar indeed 130 years later!) the 11th, 12th and 16th Light Dragoons were brigaded under Major-General Sir John Vandeleur. At Eyne during April the regimental strength was increased to 947 privates and 775 horses. The great event was the arrival of Wellington from Vienna to take command of the Army; the 11th and other regiments were delighted to be back under his command, and the Duke was cheered to the echo when he inspected Vandeleur's Brigade at Goyek on April 26, and again on May 29 when the complete British cavalry and horse artillery, 6,700 mounted men under Lieut-General the Earl of Uxbridge, were reviewed near Sandebecane by Wellington, Marshal Blücher and the Prince of Orange.

Napoleon was now faced by a four-sided alliance of England, Russia, Prussia and Austria, and with Murat defeated in Italy was compelled to fight; he crossed into Belgium from France and on June 14 forced the Prussian advance guard under Ziethen to retreat at Charleroi; two days later Blücher was attacked successfully at Ligny, and the British taken by surprise at Quatre Bras, with the Prince of Orange in command. The 11th, the nearest British cavalry to Quatre Bras, were forty-five miles distant preparing for an ordinary parade in undress uniform that morning—a measure of the success of the French surprise —and though they spurred their tired horses throughout the day only reached the battlefield of Quatre Bras when the battle was over and the French were withdrawing; the 11th could only take part in a few skirmishes.

Napoleon, however, failed to exploit the signal successes

gained by his troops at Ligny and Quatre Bras, the Prussians being allowed to withdraw unharassed and the British infantry given a precious start of eight hours' daylight to retire from Quatre Bras in a violent storm over frightful going towards Brussels and the prepared positions of the Allied Army at Waterloo. All through the wet and depressing day of June 17 Uxbridge, commanding the cavalry covering the British retreat, anticipated an all-out French attack; but although Vandeleur's 4th Cavalry and Sir Hussey Vivian's 6th Cavalry Brigades—the latter especially, being the rearguard—were repeatedly attacked during the afternoon by French cavalry as they retreated across the muddy open country and the flooded Dyle, both brigades reached their bivouacs without losing a man. Two squadrons of the 11th had moved with this body, but the other squadron, under Captain J. A. Schreiber, went with the Heavy Brigade and the main body of cavalry along the main road to Genappe and Brussels. At Genappe it beat off a strong party of French lancers.

The bivouacs round Waterloo were sodden and cheerless, for heavy rain continued throughout the night, and the only sustenance issued during the day had been a biscuit and a glass of spirits. Indeed it was a suitable occasion for splicing the mainbrace.

Sunday, June 18, 1815, that historic day, was again muddy and grey. The French numbered about 80,000, Wellington's mixed force of British, Belgians, Brunswickers and Hanoverians about 70,000. Waterloo was a battle primarily won by the British infantry, which all through the afternoon received in their squares and withstood a series of tremendous French cavalry and infantry attacks backed by accurate and intensive artillery fire. In return the British gunners fired until the last possible moment, and the infantry regiments covered themselves with glory, not least the 42nd, 79th and 92nd Highlanders of Picton's Division and the Scots and Coldstream Guards;

in the bloody, bitter, apparently never-ending struggles round Hougoumont and the farmhouse at La Haye Sainte, and against the terrible 16,000 strong attack of D'Erlon's Infantry Corps, the British infantry never broke. Napoleon could not master Wellington's men, even with Ney leading the Imperial Guards in a final desperate assault; and Blücher, whom Grouchy catastrophically had let get away from Wavre, finally and decisively arrived in the late afternoon.

The cavalry, including the 11th in three squadrons totalling 320 men in Vandeleur's 4th Cavalry Brigade, were originally drawn up in support of Wellington's left flank. After Picton's Highlanders, supported by the 1st and 28th Regiments, had repulsed D'Erlon's massive attack, however, Lord Uxbridge, unable to contain himself, launched the Royals, Greys and Inniskillings of Ponsonby's Union Brigade at the French—without orders from Wellington. At first this cavalry charge was brilliantly successful, but the horsemen pursued too far, ran into concealed French guns, were badly cut up by counter-attacking cuirassiers, and eventually retreated in disorder pursued by the French cavalry—some of which Vandeleur's brigade in turn forced back. Wellington, not unnaturally, for he knew how best to use his cavalry at the opportune time, was infuriated by this unauthorised and nearly disastrous venture of Uxbridge (who was badly wounded in the process), and sent most of the cavalry to the rear for the rest of the day until he considered it safe to loose them in pursuit of the beaten enemy. General Sir George Cooke considered it was fortunate for Uxbridge that Waterloo turned out a victory—otherwise 'he would have got into an awkward scrape for having engaged the cavalry without orders from the Duke.' Cooke also describes how Wellington greeted Cotton before the battle:

> 'General Cotton, I am glad to see you in command of the cavalry; and I wish you to bear in mind that cavalry should always be held well in hand; that your men and

horses should not be used up in wild and useless charges, but put forward when you are sure that their onset will have a decisive effect. Above all, remember that you had better not engage, as a general rule, unless you see an opportunity of attacking the French with a superior force. In Spain, the Germans, the 14th Light Dragoons, and perhaps the 12th, under Fred. Ponsonby, were the only regiments that knew their duty and did not get into scrapes of every description.'

Gronow, who appears to have been prejudiced against his cavalry colleagues, quotes Marshal Excelmann's opinion of the British cavalry, which 'struck me as remarkably instructive.' For Excelmann used to say:

'Your horses are the finest in the world, and your men ride better than any continental soldiers; with such materials, the English cavalry ought to have done more than has ever been accomplished by them on the field of battle. The great deficiency is in your officers, who have nothing to recommend them but their dash and sitting well in their saddles; indeed, as far as my experience goes, your English generals have never understood the use of cavalry: they have undoubtedly frequently misapplied this important arm of a grand army and have never, up to the battle of Waterloo, employed the mounted soldier at the proper time and in the proper place. The British cavalry officer seems to be impressed with the conviction that he can dash and ride over everything, as if the art of war were precisely the same as that of fox-hunting. I need not remind you of the charge of your two heavy brigades at Waterloo: this charge was utterly useless, and all the world knows they came upon a masked battery, which obliged a retreat, and entirely disconcerted Wellington's plans during the rest of the day.'

Brutal, perhaps exaggerated, but none the less genuine criticism; it applied less to the Peninsula veterans, however, than to others. At a critical stage of the battle near Hougoumont a Pays Bas regiment faltered, and the ever-present Tomkinson describes what happened.

> 'Major Childers, 11th Light Dragoons, and I rode up to them, encouraged them, stopped those who had moved the farthest (ten yards perhaps) out of their ranks, and whilst they were hesitating whether to retreat or continue with the column, the Duke rode up and encouraged them.'

Childers brought up his squadron of 11th and placed it in the rear of the Belgians, who then stood firm, and the Duke congratulated him and Tomkinson, who adds that if the Belgians had run away the consequences at that moment 'might have been fatal'.

When Uxbridge was wounded Vandeleur had taken his place, and in turn Colonel Sleigh assumed command of the 4th Cavalry Brigade, which, now on the right, distinguished itself in the final cavalry attack permitted by Wellington when at last the French infantry broke. Sleigh describes how, with Vivian's Brigade on his left,

> 'we took the last Battery and received their fire, which was given when my Brigade was so close that I saw the Artillerymen fire their guns. Fortunately the ground was undulating and we only lost by the fire Lt. Phelips of 11th and Hay of 16th. It was after this when continuing our advance that the 1st Hussars (King's German Legion) came up in our rear, and from its being dark were nearly in collision with us, as we knowing there was a Brigade of French Cavalry in our rear went threes about and were in the act of charging when we recognised them by their cheer. It was nearly dark and our men knew of the French Brigade being behind them.'

General Sir Evelyn Wood, V.C., writing later, says of this fifth British attack:

> 'The 16th Dragoons came on a large body of French infantry endeavouring to form square, and, charging it,

took and destroyed the entire column; while almost at the same moment the 11th Light Dragoons, further to the Westward, took a battery – the last of the French guns in position.'

The 11th, now commanded by Brevet Lieut-Colonel A. Money, also charged and broke a French infantry square, and continued the pursuit of the streaming French until Gneisenau's Prussian cavalry, fresher than the English, took over. As the fateful day ended the tired but victorious troops bivouacked where the French had camped the night before, still without food or forage, except what could be looted. Napoleon's casualties exceeded 25,000, Wellington's were 9,755 killed and wounded of his own troops; 122 French cannon and 195 gun-waggons were captured. The 11th lost one officer killed, one sergeant ten privates and seventeen horses. Four officers,* four sergeants, thirty rank and file and thirty-eight horses were wounded, one sergeant, two trumpeters, twenty rank and file and eighteen horses missing. Colonels Sleigh and Money were awarded the C.B., and Major James Bourchier and Michael Childers promoted Lieut-Colonel; all who fought at Waterloo received a silver medal and could count two years' extra service for that day, and subsequently the regiment was granted the right to bear 'Waterloo' on its guidons. Four days after the battle Napoleon abdicated for the second and final time, and on July 22 surrendered at La Rochelle to the Royal Navy. When the Emperor died at St Helena in 1821 he was almost the same age as Adolf Hitler at his death in Berlin in 1945—fifty-six—and the days of their deaths fell within a week of one another.

By the end of June Wellington's army was approaching Paris, and on July 2 the 11th crossed the Seine on pontoons and a squadron escorted the General to St Cloud. The

---

* Of these, Lieut. Robert Milligan was only eighteen; the youngest officer present was Cornet Barton Parker Browne, aged 17 years three months, who became one of the finest swordsmen in the Army and lived to the age of ninety-one.

approach march from Waterloo had revealed the brutality, looting and wanton destruction enjoyed by the Prussians, on whom Gleig, particularly shocked by the night-time looting on the field of Waterloo, commented:

> 'These men seemed to take pleasure in thus revenging themselves upon the people of a nation at the hands of whose soldiers they had in war suffered so many humiliating reverses.'

Wellington's grand entry into Paris was made on July 7, and the 11th, part of his escort, enjoyed the unusual distinction of bivouacking in the Champs Elysées; on July 24 the whole British Army in Paris was reviewed, parading between the Place Louis Quinze and Neuilly Bridge, by the Emperors of Russia and Austria and the King of Prussia. In August it was agreed that Paris should be evacuated by the victors (Louis XVIII having returned once again 'in the baggage of the allies'), but an Army of Occupation was left in France, and of this the 11th became a part, in Vandeleur's Brigade with headquarters at Neufchatel in Normandy. This army broke up in December but the 11th, now in a new brigade under Major-General Grant, remained in the occupying force, having moved from Dieppe to Villa D'Eau to the Dunkirk-Hazebrouck district (Wormhout) to St Omer to the Cambrai area (Abancourt). The regiment spent the winter of 1816-17 at Hazebrouck, where the quarters were dilapidated, the weather cold and rainy, the food bad, and fuel scarce. As usual, the British soldiers had already fraternised with the French villagers, helping in ploughing and harvesting the previous summer, so it is not surprising that Gleig relates how well planted under various tables cavalry boots were at Hazebrouck, thus making the winter more bearable; troops and peasantry 'each entertaining the other with the amusements common to their own country'—and, doubtless, with some common to both.

The summer of 1817 was spent near St Omer, the winter back at Hazebrouck; the spring of 1818 at Bailleul, the summer at Beuras, St Omer and the Valenciennes area. At some stage of the Occupation the regiment appears to have been stationed in the Low Countries for Gleig has it at Bergen in Holland in 1816, and at Moll; he also records the three days' holiday granted by Wellington after Waterloo and the generous gratuities, parts of which undoubtedly disappeared in equally generous purchases of liquor: 'The Flemings are nowise backward in their cans . . . but I suspect they never beheld such wassailing as gave a character to the three days in question.' Gleig also refers disparagingly in *The Light Dragoon* (Farmer, incidentally, had been slightly wounded at Waterloo) to an—unusually—unwelcome draft from the artillery which in his opinion lowered the standard of the regiment—'Artillery Drivers . . . in point of character, of the worst description.' These may have been part of a draft of nearly 120 men that arrived at Wormhout in May 1816.

Wellington held a final review on October 23, 1818 of the occupying forces of England, Russia, Denmark, Saxony and Hanover, before the Emperor of Russia, the King of Prussia, and the Prince of Orange. Next month the regiment, after over three and a half years on the Continent, left Calais with 476 horses and disembarked at Dover and Ramsgate on November 20 and 21, whence it marched to Canterbury. A regimental depot was established at Maidstone, the number of Other Ranks increased to 701, and the officers augmented by an extra Lieut-Colonel and eight subalterns.

Wellington's soldiers had deserved well of the nation, even if many of them were refugees from the slums, the prison cell or the bailiffs; as usual a grateful country treated them on their return to England with contemptuous indifference, much in the same way as Marlborough's men who had won half Europe had been feared and shunned.

Soldiers were still miserably underpaid, but their officers, senior and junior, especially in regiments like the 11th, were paying far more attention to the welfare of their troops than had been accorded in Marlborough's day. Despite some modern pictures of inane young officers incompetent even to run a pony club and of brutally inhuman disciplinary methods—some of which, such as flogging to excess, did exist—the proper regimental and family spirit had now flowered. As Fortescue remarks of the new generation of commanders: 'They could not, of course, sweep away insanitary barracks. There could be no arguing with legislators who provided a thousand cubic feet of space for each convict and thought three hundred sufficient for each soldier.'

But cramped space and poor English living conditions were not to worry the 11th, for almost immediately they were ordered to India for the first time. On February 9, 1819, two East Indiamen bore them down the Thames headed for an unknown sub-continent where they were to spend the next nineteen years. With them went Private Farmer, determined in spite of periodical 'bad drafts' which met with his disapproval to maintain the quality of the regiment he loved: 'For we—the old hands of the 11th —were much attached both to our regiment, considered as a body, and to the officers and non-commissioned officers belonging to it.'

Chapter 5
To the East

INDIA was a completely new experience. For young soldiers there was certainly the fascination of a mysterious and colourful new scene, but there was also dirt, poverty, boredom, and worst of all disease. In its first twenty years in India the regiment was to lose as many if not more men from epidemics and illness as it had suffered casualties in the Peninsula and Waterloo campaigns. It got off to a bad start: on the slow passage up the Ganges from Calcutta to Cawnpore between the end of July and October 1819 no less than twenty-five men died, and by September 1820 it had lost three officers and 166 men, mostly from fever; in 1821 another officer and 28 men died. Regimental strength in December 1819 was 726 men, and 654 horses (taken over from the 21st Light Dragoons, along with ninety-two volunteers), but because of the Government system of deductions in horse purchases from native dealers many of the mounts were inferior to English standards. When Brevet-Colonel Sleigh left the regiment in 1821 after twenty-six years' service he noted the following changes during his time: Officers 231, other ranks 2,506, horses 4,449.

For most of this period the regiment was stationed either at Cawnpore or Meerut, where its duties were not arduous; but in 1825 it saw fierce action at the siege of Bhurtpore, a strongly fortified town eight miles in circumference set in a sun-baked plain thirty miles west of Agra. This great Jat fortress was considered impregnable, since in 1805 Lord Lake had failed to capture it; the bones of thousands of his

British troops had been left to whiten among the bastions and Bhurtpore had become a haven for malcontents who objected to British rule in Upper India. Now with the death of the Rajah came trouble, since his nephew Doorjun Saul seized the throne and Bhurtpore itself to the exclusion of the rightful heir, who was under British protection.

When peaceful negotiations availed nothing, a large British and native force of 30,000 men with a strong siege train was assembled, over which Sir Stapleton Cotton, now Lord Combermere, was put in command. An 11th Light Dragoon, Brigadier-General Sleigh, commanded the cavalry, Lieut-Colonel M. Childers one of the two cavalry brigades, in which was the regiment, now commanded by Major Bellingham Smith.

The strength of the garrison is given by General Cust as only 2,500 men, but Gleig says there were about 15,000 men in Bhurtpore. Certainly they were well armed, provided, and protected by a strong natural position extensively fortified; the garrison came largely from the most warlike races of India, and disposed of a good number of cannon, amongst which was an enormous ceremonial gun known as 'Sweet-Lips' which was discharged 'only at stated seasons' for particular effect on the Rajah's behalf. (It now ornaments the parade by St James's Park.) Gleig comments: 'It is an extremely beautiful piece of mechanism, but considered as a weapon of war, was perfectly useless.' The town, with its nine main gates, fort and citadel, all with very solid walls, stood in the very heart of an enormous wood some 5–6,000 yards from the palace at its fringes, and through this wood ran many roads and rides. Although far superior in numbers, the bulk of the British force was made up of sepoys (Gleig says there were only three British regiments present, presumably referring to the cavalry only, which contained the 16th Lancers, 11th Light Dragoons, and four Troops R.H.A.; the British infantry certainly included the 14th and 59th Regiments).

At all events, the Bhurtpore garrison suffered from over-arrogance:

> 'Nothing could exceed the carelessness, or the misplaced confidence of its ruler and his troops. Though they must have been long aware of our hostile designs, they took no precautions whatever to defeat or even to retard their accomplishment.' (Gleig.)

Action started on December 10 and there was a good deal of building of siege works and artillery duelling; the 11th were divided between mounted patrols and work in the trenches, and on one occasion Lieut. Wymer cut off an enemy raiding party from the fort in deep jungle much better known to them than to him. The besiegers completed their first parallel just before Christmas, but though Bhurtpore was soon reduced to ruins by the resulting artillery fire the mud walls surrounding it proved too tough to batter. Various mining and counter-mining operations then went on, through one of which the garrison scored their only real success by blowing up 20,000 lbs. of British gunpowder, but by January 17 when a huge mine had been completed on the north-east angle of the works Combermere decided it was time for storming operations. Major Smith, two subalterns and eighty men of the 11th volunteered—with a like number of the 16th Lancers—to take part.

When the immense mine was exploded at eight a.m. next day, an enormous breach was made and scores of large stones and masses of earth were flung through the air, burying some of the defenders but also through miscalculation killing and wounding a number of the assault party. Resistance was fierce, with volleys of grape, round shot and musketry poured on the attackers, but in two hours the ramparts had been cleared and by late afternoon the gates of the citadel of Bhurtpore were captured.

Doorjun Saul, with 160 chosen horsemen, tried a break-out, but this had been foreseen by Sleigh, and the ruler and his party were taken by the 8th Indian Light Cavalry. While a detachment of the 11th was pursuing some twenty or thirty horsemen also escaping from the fort Pte Farmer saved Sergeant Waldron by cutting down a ferocious-looking Rajput aiming an English-made carbine at him. Another prisoner taken (who was at once hung) was Bombardier Herbert, who had deserted on Christmas Day and subsequently directed the enemy's artillery fire to great effect, on one occasion nearly shooting Lord Combermere.

The Bhurtpore garrison suffered losses of some 4,000, mostly in the final explosion and assault, the British regiments sixty-one killed and 283 wounded, the Indian regiments forty-two killed and 183 wounded. The 11th had two men killed, one officer and twelve men wounded, four horses killed and twenty-two wounded.

The Bhurtpore affair had riveted India's and indeed outside attention, for it came at a critical time when Britain was engaged in the Burmese War and only a spark was needed to set much of India aflame in revolt. Combermere's efficient action damped out the more ardent revolutionary fires and bolstered the East India Company. In his despatches he commended the important services of Sleigh and Childers (who was later made C.B.) and also the 'conspicuous gallantry on numerous occasions' of Major Nicholas Brutton, an old and much-wounded India hand who had come to the regiment from the 8th Light Dragoons and was later to command it. Another good officer who was at Bhurtpore was Captain James Rotton, commissioned a Cornet at sixteen and blooded in the Peninsula at seventeen. But of all the Bhurtpore experiences perhaps the pleasantest concerned Major Bellingham Smith, who came across a small Indian child weeping bitterly at the side of his dead father. He promised through his Indian servant to be a protector to the child and Gleig recounts: 'He faithfully redeemed the pledge.'

Immense booty was taken at Bhurtpore, and every private soldier got 40–50 rupees in prize money; this did not seem much when contrasted with Combermere's share (Rs. 595,398) or even that of a Major (Rs. 9,500).

In 1830 the regiment resumed scarlet clothing, its coats having been blue since 1784. During this first Indian tour the inspection reports were uniformly good, but as in other regiments there were outbreaks of excessive drinking and quarrels that led to bloodshed; almost invariably the cause was boredom or inactivity, though Gleig summed up that life in India for the common soldier was 'not . . . entirely a strange intermixture of military duty, and dissipation, and sheer idleness.' In 1835 Pte Farmer applied for his discharge, which he received at Chatham the following year: his pension after twenty-eight years' service was the glorious sum of—10*d.* per day! Farmer did not consider this was 'too great' after all his efforts, and there is a sad note in his comment, 'I am by far too much worn out to add to it greatly by personal exertion.'

It was not only the rank and file whose services often passed meagrely rewarded. When Colonel Brutton retired after over forty-two years' service in the Army, most of it in India where he had greatly distinguished himself in numerous campaigns and many times nearly lost his life, he had one campaign medal and was granted a pension of £100 per year for his wounds. The officers clubbed together to give him a handsome piece of plate to show their gratitude for 'the uniform kindness' they had been shewn by him while he was their commanding officer. This kindly and able Devonian was to be succeeded by a very different character, hardly remarkable for his affability, especially towards junior officers, for in 1836 Lieut-Colonel Lord Brudenell (who had been ousted from command of the 15th Hussars a few years before) was gazetted to command of the 11th *vice* Lieut-Colonel Michael Childers, and in October 1837 he took over the regiment at Cawnpore from

Colonel Brutton. Despite mutual expressions of esteem, it must have been hard for the old professional to hand over to the young, untried but arrogant amateur. A few months later the regiment left Calcutta for England, and it was another tribute to the enfeebling climate of India that of 344 N.C.O.s and men who returned, 120 were invalided; another 158 had elected to stay in India, 110 going to the 3rd Light Dragoons and forty-eight to the 16th Lancers.

It was in a sense the end of an era, the era of Wellington and his great victories, for most of the soldiers in the regiment who had campaigned with him had now passed on to promotion, death, or retirement. Wellington, still the supreme arbiter of military affairs, had begun his great career in India as a 'Sepoy General'. It was also the beginning of the Victorian age—and for the regiment, more important, what can only be called the startling, extraordinary and tragi-comic Cardigan era.

Chapter 6

The Cardigan Era

'It may be necessary to observe, that among continental troops, no man who has begun in the cavalry and above all in the Hussars, ever dreams of enrolling himself in a regiment of infantry. Such a step would be accounted a degradation, and as hussar regiments in general are dressed with very great magnificence, the individuals belonging to them acquire an esprit de corps, such as you will scarcely find in any other armed body.'
(G. R. Gleig, *'The Hussar'*, Henry Colburn, London, 1837.)

THE 11th Hussars, like some of its favourite fellow regiments such as the 12th Lancers, the Royals, the 5th Inniskilling Dragoon Guards, and the 8th Hussars, has never been a particularly rich regiment, though it has had its share of affluent aristocrats. But neither money nor birth, one is happy to state, from its very beginnings right up to the present time have ever provided the key to success within the regiment, bearing in mind that like every other unit it was subject to and sometimes suffered from the quirks and unfairnesses of the purchase system which endured for so many years through the eighteenth and nineteenth centuries. But even in its early days it was noticeable that many who came to the 11th from other regiments stayed, if they could, with it for the rest of their time, both officers and men, and others who had to leave often came back if human ingenuity could manage it; even in the early nineteenth century men rose from the ranks to be officers. Both these characteristics were never more noticeable than during Hitler's war, and one likes to think that they show an appreciation of real values, of the comradeship, humour, ability, sense of family and professionalism that constitute what is generally known as the regimental spirit.

By the time they left India in 1838 the 11th Light Dragoons were as professional as they had emerged from

Army Museums Ogilby Trust
A sergeant of the 11th Hussars during the 'Cardigan Era'.

the Napoleonic wars, and in some ways perhaps more so, despite the few opportunities for action. India has always been a great training ground, and men like Sleigh, Childers, Bellingham Smith and Brutton were experienced and also kindly officers of high ability. Coming as most of the regimental officers did from the landed gentry of England and Ireland, rather than from the gilded aristocracy, they had a hard core of common sense, and even in the age of privilege and extraordinary rigidity of social barriers often became as devoted to their soldiers as to their tenants at home.

Thus the 11th, a happy regiment, must have watched with alarm and misgivings the battle that broke out when Lord Brudenell was appointed their next commanding officer in March 1836—having reputedly paid over £40,000 for the purchase of the regiment. There must have been many who hoped that this appointment would never come about, for the news of Brudenell's extraordinary and apparently almost demented behaviour as Commanding Officer of the 15th Hussars, another notably efficient regiment, had percolated even to distant India. It had culminated in 1834 in the acquittal by general court-martial of the popular and respected Captain Augustus Wathen, who had been endlessly victimised and insulted on pettifogging grounds by Brudenell, and the removal from command of the 15th of the latter.* James Thomas Brudenell, however, believed firmly in the divine right of Cardigans as well as kings; he refused to be, let alone feel, discredited, and no doubt ever crossed his mind that he might not be in the right. Utterly self-absorbed, Brudenell was also fearless, handsome, rich, spoiled, ambitious militarily but quite without operational experience, though he had spent six years in the 8th Hussars; he was a fine horseman, but at Deene Park, seat of his amiable father the sixth Earl of Cardigan, he had

---

* These findings were read at the head of every regiment in William IV's service.

grown up surrounded by women, mostly his many sisters, which had encouraged him to be vain, domineering and undisciplined. At times his temper was almost uncontrollable. His tastes were Philistine but expensive, and he had been a prominent member of the Bullingdon set of the day during the two privileged years he had spent at Christ Church (going down from Oxford without taking a degree). Withal Brudenell, essentially a spoiled and rather dangerous child, retained a childlike simplicity; and although during his tenure of the 15th he had introduced the revolting habit of having the conversation of his officers in the orderly room taken down without their knowledge, he was normally straightforward and guileless. Later, indeed, the historian A. W. Kinglake was to say that Cardigan was as innocent as a horse. It would be a mistake, however, to dismiss him as a fool. He lacked judgement and common sense, often behaved ludicrously if not crassly, but above all his failure was an incapacity or unwillingness to understand and sympathise with the points of view and feelings of other people; he was the very opposite of the 18th Century Man of Feeling of Henry Mackenzie. Yet he had some of the native shrewdness of a fox and a certain bluff geniality. Like many people who will not be crossed he saw offence where none was or was meant. And like a previous Lord Cardigan described in a letter of Mr. E. Lewis to Lord Bruce in 1736, he tended 'to look upon every man that throws difficulties in his way in an adverse light. . . .'

Such, then, was the officer with whom the 11th were to be saddled. It had taken a great deal of lobbying and thick-skinned solicitation of the great, including Lord Melbourne (the Prime Minister), Lord John Russell, the Commander-in-Chief (Lord 'Daddy' Hill), and even the Duke of Wellington, still final arbiter on military matters, to ensure Brudenell a return from the wilderness on suffrance; finally King William IV, after pressure from his wife Queen Adelaide, to whose Chamberlain one of

Cardigan's sisters was married, and a sentimental interview with the sixth Earl of Cardigan, had agreed against his better judgement to reinstate Lord Brudenell.

The Press and public, however, were not so easily satisfied, and a storm of indignation followed, culminating in a motion brought in the House of Commons by Sir William Molesworth demanding a Select Committee to inquire into the new appointment. Molesworth, an unpopular Radical, was attacking not only Brudenell but Lord Hill, whose judgement he considered, not unjustly, had been warped by courtly influence and intrigue. Reiterating the background and verdict of the Wathen court-martial he broadly took the line that Brudenell's appointment would be an insult to the 11th: 'They will murmur, and most justly— loud will be their indignation.' Molesworth contended that Brudenell was not a fit or proper person to command such a distinguished body:

> 'A more gallant regiment does not exist in his Majesty's service, nor one that has better served its country – in Egypt, in the Peninsula, at Waterloo, with the army of occupation in France, thence – seventeen years ago, removed to India, at Bhurtpore, and elsewhere, it has distinguished itself: some of its officers have been nearly as many years in the army as the Noble Lord has lived years in this world. The two majors have served with this regiment since the years 1806 and 1811 [presumably John Jenkins and James Rotton, who joined in 1807 and 1811 respectively]. With what feelings will they view the advancement over their heads of this young officer, who has never heard the sound of a musket, except in the mimic combats of a review. . . .'
> (Commons Debate of May 3, 1836.)

Unexpectedly Brudenell employed a dignified and restrained tone in his own speech of self-justification to the House, at the end of which he was loudly cheered. He disclosed that he had requested a court-martial of himself, and apologised to the majority for boring them with military details of no interest to them, 'but of the greatest importance to myself, as being intimately connected with my honour and character, which are as dear to me as life itself.' This went down well. He also quoted a letter to himself from Major-General Sir Frederick Ponsonby, a respected figure, opining that he had been treated with great severity, and that if he was not permitted to exchange into the 11th Dragoons

> 'you will be treated with positive injustice. I gave you this opinion in December last, and I am strengthened in the belief that it is correct, from having conversed with many officers of all classes in the army, and I have found scarcely anyone who has differed with me.'

Brudenell also produced warm testimonials from other distinguished officers, including one from Major-General Sir Edward Blakeney in Ireland (where the 15th had been stationed) that strains credulity now as then when Blakeney says:

> 'In field movements, and the ready application of cavalry, I considered you one of the most intelligent officers that served under my orders.'

The House was already won over when Joseph Hume, another leading Radical, supporting Molesworth attacked by clear implication not only Lord Hill, for pursuing a course which had 'set discipline at defiance', but the Court. As Hume was loathed by the Army authorities for his efforts to have flogging abolished his speech provoked ribald outbursts rather than sympathy. The day was finally swung by Major-General Sir Henry Hardinge, a close friend of Wellington's, who warmly supported Brudenell, dismissing the Wathen verdict as 'an erroneous decision' and the court-martial's animadversions on Lord Brudenell as mistaken. Molesworth then tried unsuccessfully to withdraw his motion, but it was voted on and defeated by 322 to 42 votes—Gladstone being one of the Noes.

At first all went well with the confirmed appointment, if mainly because Brudenell, who succeeded his father as Earl of Cardigan in August 1837, was hardly ever with the regiment until it returned to England. This acceptable state of things was not to last. Cardigan began to find fault with everything, from horse furniture to uniforms to drill to riding performance to behaviour in the Mess; above all he now made clear his contempt for anything to do with Indian service and thus for the old Indian hands—the majority, in fact, of his officers. Some who found his attitude impossible left the regiment, most writhed inwardly but soldiered on. Although in August 1838 the establishment was down to about 350 all ranks and only six Troops, Cardigan managed to hold fifty-four courts-martial before the end of the year, and according to Mrs. Woodham-Smith 'desertions were numerous, and Canterbury jail became so filled with soldiers of the 11th that it was said to have become their regimental barracks.' Against this, however, must be put the exchange of complimentary letters two years later, when the regiment left Canterbury for Brighton, between the Seventh Earl and the City Corporation, which, unless it was being sarcastic, was sorry to see them go ('Deeply do we regret your departure from our ancient city . . .') and referred to 'the noble, courteous, and gentlemanly bearing of your Lordship and the officers under your command, as well also of the meritorious and respectful deportment of the men. . . .' In fact the 11th had always had a particularly good reputation for behaviour in barracks or cantonment.

In 1839 various innovations in equipment were made, the most important of which was the issue of new Victoria percussion carbines. Whatever his shortcomings of personality, Cardigan was undoubtedly efficient, and it is too facile to dismiss him as a figurehead, or to say, as does Sir John Fortescue, 'Cardigan seems only to have been a better kind of sergeant-major.' Besides this, he now had a great deal of money, with an income of £40,000 a year, a quarter of which he was reputed to spend on the 11th. George Ryan relates how Cardigan, dissatisfied with the standard of mounts, waited until the Horse Guards ordered a remount and then 'most spiritedly' added £10 to the price allowed by the Government for each horse. 'By this act of munificence, the 11th Hussars became the best mounted regiment in the service.'

Certainly the Duke of Cambridge, Lord Hill, Lieut-General Sir Hussey Vivian, and even the Duke of Wellington all expressed unusually warm approval of the efficiency

Army Museums Ogilby Trust
A sergeant of the 11th Hussars holding 'Mercury,' Prince Albert's horse. Reproduced by gracious permission of Her Majesty the Queen.

and turn-out of the regiment in 1839. Cardigan was getting some return for his money.

Eighteen hundred and forty was a memorable year, when Queen Victoria married Prince Albert of Saxe-Coburg and as a direct result the title of the regiment was changed. In turn squadrons of the 11th Light Dragoons escorted Prince Albert on February 7 and 8 from Dover to Canterbury and along the London road on the way to his marriage. On March 14 Cardigan received a letter from Lieut-General Macdonald, the Adjutant-General, which read:
'My Lord,

I have the honour to acquaint you, by the direction of the General Commanding in Chief, that Her Majesty has been graciously pleased to direct that the ELEVENTH Regiment of Light Dragoons shall be armed, clothed and equipped as Hussars and styled the "ELEVENTH, or PRINCE ALBERT'S OWN HUSSARS".'

Consequently the uniform was changed from a shako, scarlet light dragoon jacket and blue overalls to a busby with crimson and white plume, blue jacket, pelisse and crimson overalls, with gold lace fittings, and the shabracque became crimson instead of blue. The crimson trousers were taken from the Saxe-Coburg livery, and the 11th later became the only Hussar regiment—as it still is—to wear different coloured trousers and tunics with service and walking-out dress. Prince Albert had been so impressed by his escorts that he asked for the regiment to be made his own, a wish that was granted when the Queen appointed him Colonel on April 30, 1840. The 11th Hussars took his crest and motto as its badge—a pillar charged with the arms of Saxony rising out of the ducal coronet, and also crowned with a like coronet out of which rise five feathers, under the whole a scroll containing the motto '*Treu und Fest*' ('True and Faithful'). Apart from '*Ich Dien*' on the Prince of Wales's plume this is the only German motto borne in the British Army. It is still on the cap-badge and

Army Museums Ogilby Trust
Prince Albert in the uniform of the Colonel, the 11th Hussars.

'collar dogs', and on arm badges worn by Sergeants and senior N.C.O.s. Old coloured prints of the 1840s show that the new uniforms, particularly full dress, were probably the most gorgeous in the Army—at a time when gorgeous uniforms were commonplace. *The Times* was moved to write:

> 'The brevity of their jackets, the irrationality of their headgear, the incredible tightness of their cherry coloured pants, altogether defy description: they must be seen to be appreciated.'

Seen they were, for even when the regiment was at Canterbury or Brighton it was Cardigan's habit to give specially picked men a one day pass and five shillings, with instructions to post themselves at strategic points in London such as St. James's, Piccadilly, and Hyde Park Corner, where in all their glory they could salute him as he passed by. Despite the outward show, however, all was far from well in the regiment; the arrogance, unforgiving obstinacy (and unfortunate taste for litigation that often goes with those qualities) showed by the Colonel did not modify. Nor would he adapt his absurd attitude towards the 'Indian' officers. Martinets were not uncommon in the army, then as now, but there was nothing of fairness in Cardigan's persecution of some of his officers. The first great row started on May 18, 1840, when the Inspector-General of Cavalry, Major-General Sleigh, an old 11th Light Dragoon, was dining in the Mess at Canterbury. It was the famous or infamous 'Black Bottle' incident. Cardigan had ordered that nothing but champagne should be drunk, but one of Sleigh's staff asked Captain John Reynolds if he could have Moselle instead, and this was served direct in its bottle. Cardigan, seeing the bottle on the table, flew into a fury and accused Reynolds of drinking porter, which had been much favoured in India. Even when the explanation was given Cardigan refused to be satisfied, barking that gentlemen decanted their wine; and next day he dressed down Reynolds in front of other officers, suggested, amid a blanket condemnation of the indiscipline of those who had served in India, that he leave the regiment, thoroughly abused him, and finally placed him under close arrest. In June Sleigh appeared once more, summoned all the officers of the regiment, and read a letter from the Horse Guards that condemned Reynolds as strongly as possible and backed up Cardigan—though there had been no enquiry of any sort. Reynolds then asked for a court-martial, a request that enraged Sleigh, who told him that he had 'forfeited the sympathy of very officer of rank in the service' and declared that the matter was finished and done with. Thus Reynolds stayed under an unpleasant cloud, but not for long, since his position was clearly impossible. He applied for leave to study at the senior department of the Royal Military College (later the Staff College), but Cardigan spitefully refused so Reynolds decided to sell out. The affair did not end here, however, for news of it had leaked out, the newspapers and music-halls were full of jokes at the expense of the regiment and Cardigan, and Reynolds's guardian was deluging Lord Hill and the Press with protests, most of which were printed. All of this was bad for the regiment, if not for the Army as a whole, and when Cardigan reprimanded one of his soldiers for hitting a guardsman in the street the private pleaded, 'My Lord, he called me a black bottle.' Hill was forced by the manifest injustice of the case to write to Cardigan very sharply, and in January 1841 deputed Lord Fitzroy Somerset (later Lord Raglan) to persuade Reynolds not to sell out; Somerset was a friend of Cardigan, and with his considerable charm urged Reynolds to withdraw his resignation for the good of the service. Reynolds finally agreed on his own terms, 'the first being that he should never again be required, even for a single day to serve under the Earl of Cardigan', also the withdrawal of Sleigh's censure, six months' leave and permission to study at the R.M.A. The Press received the news with acclamation.

All was by no means healed or well, however, for already Cardigan had been victimising another Captain Reynolds, Richard, with long Indian service, whom he reprimanded in front of the regiment and refused any sort of leave. In August 1840 Cardigan gave one of his lavish parties at Brunswick House, Brighton, and at this glittering affair from which both Captain Reynolds were conspicuously absent a young lady asked her host why they were not there. Cardigan, before a number of other guests, replied,

'As long as I live they shall never enter my house!' and next day Richard Reynolds, not unnaturally having heard of this disparagement, wrote his Commanding Officer a letter of protest. Again he was blackguarded in front of the regiment, again the insults were tied in with references to 'Indian' officers and their inability to behave. This was too much for many known and unknown officers who had served in India, whose protests re-echoed not only through Whitehall but in the Press, *The Times* printing a letter which demanded that the C.-in-C. India on behalf of all officers who had served there should make Cardigan apologise.

Unfortunately Reynolds now wrote a rash letter to Cardigan, evidently feeling that his only recourse was to call out his tormentor. But it was clearly a breach of discipline to write, as he did:

> 'Your Lordship's reputation as a professed duellist . . . does not admit of your privately offering insult to me, and then screening yourself under the cloak of Commanding Officer and I must be allowed to tell your Lordship that it would far better become you to select a man whose hands are untied for the object of your Lordship's vindictive reproaches, or to act as many a more gallant fellow than yourself has done, and waive that rank which your wealth and Earldom alone entitle you to hold.'

As a result of this pardonable but foolish outburst Richard Reynolds was court-martialed, the trial lasting a fortnight; insomuch as Reynolds had clearly flouted an Article of War by provoking his Commanding Officer to fight a duel, the outcome was really never in doubt, though public sympathy was very much on his side, and on October 19 it was announced that he had been cashiered. Eighteen months later, against the wishes of Wellington, the story had a happy ending when he was gazetted Captain in the 9th Lancers, the court-martial's verdict being rescinded. What was not generally known, according to Ryan, was that

Reynolds had received his original Cornetcy in the regiment in 1826 through Cardigan's influence, though as Cardigan was then only a subaltern in the 8th Hussars and had no evident connection with the 11th this may not be so.

Before the court-martial, however, further fuel had been added to the flames by the publication in the *Morning Chronicle* of two anonymous letters severely criticising Cardigan's conduct to his officers, specifically the two Captain Reynolds. These, it was no secret, were written by Captain Harvey Tuckett, a popular Bhurtpore veteran who had left the 11th recently. From Cardigan's viewpoint this was more sinister, since Tuckett not only had inside knowledge but could not be got at. So he sent Captain Douglas, one of his own recent imports into the regiment, to see Tuckett and obtain an apology. This Tuckett refused, with the result that on September 12 a duel was fought on Wimbledon Common between the two, Cardigan's pistol severely wounding Tuckett at the second exchange.

Cardigan was now at the height of his unpopularity, the object not merely of Army or local but of national odium, though within the regiment he had his supporters such as Douglas and the Adjutant, Captain Inigo Jones. Richard Reynolds generously acknowledged his fault in public and asked that the curious demonstrations against Cardigan should cease (efforts were being made to petition the Queen to remove him from his second command). Booed at Brighton, hissed and jeered at Drury Lane, everywhere he appeared in public the cause of uproar and shouts of 'Reynolds!' 'Shame!' and 'Black Bottle!' Cardigan, true to form, remained impervious and indeed dignified in the face of execration. As usual, he was utterly convinced that everyone but him was out of step. No theatrical performance at which he put in an appearance could take place until he had left, and there was such an outburst from the audience at the Brighton Theatre on November 2 that *The Times* commented: 'Such effrontery, such a defiance of public opinion was not looked for even from the commander of the 11th Hussars.'

All this would have been amusing enough but for the potentially disastrous and already harmful effects upon the regiment itself and the reputation of senior officers in general. No wonder that Hill and Wellington were alarmed and infuriated as they saw the effects on discipline throughout the whole Army. The latter wrote feelingly to the C-in-C:

> 'At a moment at which the Queen's Service may require the greatest professional effort on the part of every organised military body, it may be found that this fine corps [the 11th] in the highest order and discipline, the conduct of its non-commissioned officers and men excellent, is unfit to perform the service required from it on account of these party disputes between the officers and the Commanding Officer upon petty trifles. . . .'

Wellington suggested that Hill should send a General to make clear to the regiment that such disputes with their deleterious effect upon efficiency could not be tolerated. This task was imposed by Hill, who warmly agreed with the Duke, on his Adjutant-General, Sir John Macdonald, who held a meeting of all officers behind locked doors at Brighton on October 22. Macdonald told them that Hill was adamant against hearing further complaints about Cardigan's conduct in the past, though he would enquire into any future conduct that warranted it. Sir John then urged the development of friendly feeling all round, and addressing Cardigan very distinctly relayed to him Hill's desire that he would in future exercise moderation and forbearance in his command, adding, according to *The Times*, which had apparently been briefed:

> 'It was Lord Hill's opinion that the numerous complaints which had been made to him as Commander-in-Chief would never have occurred if the Lieutenant-

Colonel of the 11th Hussars had evinced the proper degree of temper and discretion in the exercise of his command.'

Cardigan was visibly shattered by this rebuke.

There remained the possibility that he might be convicted of the criminal charge of duelling, which was a capital offence if a wound was inflicted, and even without could lead in theory to three years' hard labour at the treadmill or long-term transportation. In practice, as duelling was the perquisite generally of aristocrats, such penalties hardly ever were paid. In Cardigan's case, despite Tuckett's wounding, the Crown dealt only with intent, and though the facts of the affair were hardly in dispute, when Cardigan was tried by his peers in the House of Lords on February 16, 1841, the Lord Chief Justice, Lord Denman, connived at the defence's sufficiently flimsy claim that identification of Captain Tuckett on Wimbledon Common had not been proved and virtually directed an acquittal. The peers acquitted Cardigan *nem. diss.*, and next month his second, Douglas, who as a commoner appeared at the Old Bailey as an accessory to attempted murder, was also acquitted. Cardigan celebrated his acquittal by another visit to Drury Lane; another riot occurred.

It might be thought that by now enough notoriety and wrangling had beset the regiment, and that with its removal nearer London to Hounslow, with detachments at Hampton Court, Kensington and Sandhurst, on 'Queen's duty' on April 1, 1841, things would quieten down. Not a bit of it. On Easter Sunday, April 11, with an ineptitude that even for him was clumsy, directly after church parade in the riding-school Cardigan reassembled the regiment there to witness the flogging of a private who had committed some offence. Even at this remove one can imagine the hullabaloo that was raised in the Press, well before the days of the *News of the World* or *The People*. Indignant questions were asked in the Commons (some by Joseph Hume), and despite a special Order of the Day issued by Hill rebuking

Cardigan's conduct very plainly, there was a real danger in the Government's eyes that pent-up public feeling against Cardigan might now explode with dire results to the Army and themselves. It was no joking matter when the conduct of one regimental commander required a special Cabinet meeting—at which it was decided that Cardigan must go. Lord Melbourne, the Prime Minister, told Hill he must advise the Queen to that effect, and told the Queen that Hill, though 'deeply chagrined and annoyed', was considering the matter.

In his letter of April 24, 1841 to Queen Victoria Melbourne writes:

> 'We have had under our consideration at the Cabinet the unfortunate subject of the conduct of Lord Cardigan. The public feeling upon it is very strong, and it is almost certain that a Motion will be made in the House of Commons for an Address praying your Majesty to remove him from the command of his regiment. Such a Motion, if made, there is very little chance of resisting with success. . . .

'The danger of the whole of Lord Cardigan's proceedings has been lest a precedent of this nature shall arise out of them', continued Melbourne, in a reference to the bringing of Army discipline and internal matters under the inspection and control of the Commons because of complaints by officers and even by private men.

> 'The opinion of the Duke [Wellington] is that the Punishment on Sunday was a great impropriety and indiscretion upon the part of Lord Cardigan, but not a Military offence, nor a breach of the Mutiny Act or the Articles of War; that it called for the censure of the Commander-in-Chief, which censure was pronounced by the General Order upon which the Duke was consulted before it was issued, and that according to the usage of the Service no further step can be taken by the Military Authorities. This opinion Lord Melbourne will submit today to the Cabinet Ministers.

11 H.—4*

'Lord Melbourne perceives that he has unintentionally written upon two sheets of paper, which he hopes will cause your Majesty no inconvenience.'

Wellington himself wrote to Lady Wilton from London on April 26, 1841:

'I have been for the last days up to the Ears as usual in Lord Cardigan's last indiscretion. I had first to settle it for Lord Hill to draw the order. I have since settled it with the Government . . . this is no small Affair settled.'

As Sir George Trevelyan observed in his *Life and Letters of Lord Macaulay* (then Secretary at War) the list of Cardigan's achievements during a single twelvemonth was remarkable; as the Government was a Whig one and Cardigan a diehard Tory, it was only the great authority of Wellington and his insistence that the Commons should not interfere with the Army that saved Cardigan.

Of this Cardigan appeared unaware, though he realised that the Sunday flogging had drawn down his superiors' disapproval. But throughout the fifteen years or so of his regiment's English and Irish duty before the Crimea he continued to hold one ace of inestimable value—the goodwill and even approval of the Queen. The reason for this is not hard to find, for apart from the high birth and traditional access to the Court of his family, heavy enough weapons in Victorian England, the 11th was now *Albert's* regiment, the Queen was flattered and admired its panache and smartness on the various royal occasions at which it paraded or furnished escorts, and frequently expressed her 'approbation' of its performance of its duties; Prince Albert did the same when inspecting its field movements, and when on April 20 at a review on, ironically enough, Wimbledon Common, he rode past at the head of the 11th as its Colonel for the last time and gave her the salute, how proud Victoria was and how pleasedly she conversed with the Duke of Wellington on the fine sight. Five days later Albert removed to the Colonelcy of the Scots Fusilier Guards, but Her Majesty commanded that the 11th should retain the designation *'The Prince Albert's Own Hussars'*.

At this time Cardigan was assiduously lobbying the new Prime Minister for appointment to the Garter, in which Order a vacancy had occurred. Such soliciting was almost without precedent, though if Melbourne's reason for liking the Garter—that there was no damned merit about it—had prevailed Cardigan might well have been the likeliest contender. Peel, whose nature was not one to approve Cardigan, informed the Queen on March 20, 1842 that he had never received a direct application on the subject of the Garter, except from the Duke of Buckingham and now from Cardigan. The Queen, possibly euphoric after the Wimbledon Common review of April 20, wrote on that very day to Peel, enclosing Albert's acceptance of the Scots Guards. She told her Prime Minister:

> 'At the same time, both the Prince and Queen feel much regret at the Prince's leaving the 11th, which is, if possible, enhanced by seeing the Regiment out to-day, which is in beautiful order. It was, besides, the Regiment which escorted the Prince from Dover to Canterbury on his arrival in England in February '40. The Queen fears, indeed knows, that Lord Cardigan will be deeply mortified at the Prince's having to leave the Regiment, and that it will have the effect of appearing like another slight to him; therefore, the Queen much wishes that at some fit opportunity a mark of favour should be bestowed upon him. . . .
>
> 'The Queen hopes Sir Robert will think of this.'

But Peel made no recommendation, though Cardigan continued to press for the Garter whenever vacancies arose. He was never to receive it.

By the time the regiment was posted to Ireland in April 1843 there was hardly an officer left who had served in India. Nevertheless Cardigan continued to find causes for trouble, to abuse his officers, and on one occasion put the senior subaltern under arrest for being ten minutes late for stables. His habit of dressing down officers in front of the whole regiment eventually provoked a round robin which was sent to Wellington, now again C-in-C, who threatened in his reply to disperse all the officers throughout various other regiments if this dissension did not cease. The Duke also demanded that copies of any future complaints from officers to Cardigan and of Cardigan's replies should be 'submitted to the General Officer who should be so unfortunate as to have the Regiment under his command'. The Duke's patience was nearly exhausted, and all the 11th—except Cardigan—took this hard. When Cardigan refused an extension of compassionate leave to one of his best officers, Captain (later Lieut-General) William Forrest, leader of 'The Rebels' in the 11th, with whom he had previously quarrelled, and whose wife had had a difficult birth, Wellington was again brought into the dispute, which had become violent. This time his temper was strained to the limit and he told Cardigan off in downright terms, since Forrest was clearly in the right, adding for good measure that if any other regiment in his army could not be commanded 'without such voluminous correspondence and such futile details' he would need an additional staff.

> 'The details of the foolish quarrels of the officers of the 11th among themselves had gone to such an extent that if they continued the Duke might think it necessary to submit to Her Majesty some plan to relieve the Department from an intolerable annoyance.'

In 1845 the Duke's displeasure was caused by a totally different event, yet one in which, it now seems inevitable, the protagonist was an 11th Hussar, Captain (later Lieut-Colonel) Charles Ibbetson. Wellington was moved to write on November 17, 1845, to his confidante Lady Wilton:

> 'I see that you had not yet received the Report of the Horrible event in the Jersey family, the elopement of

Lady Adela from Brighton with a Captain Ibbetson of the 11th Hussars, to whom she was married at Gretna Green!

'It is useless my writing you an account of the event, of which you will find the details in all the Newspapers. The father, Mother and whole family are in Despair, and very reasonably. I think it is altogether the worst Affair of the Kind that has ever come under my observation. Neither Lord nor Lady Jersey nor Clementina have had the slightest knowledge of the manner in which she became acquainted with this gentleman. I asked if he had danced with her at Almack's or anywhere? They all answer no! . . .

'It appears that he lived in a House not far from Lady Jersey's, and that he was seen constantly looking through a sea Spying Glass fixed upon Lady Jersey's House; that he made Signals from His window and Verandah. . . .

'In short it is altogether the most astounding event!'

Captain Ibbetson, clearly a believer in the axiom that time spent in reconnaissance is seldom wasted, married the seventeen-year-old Lady Adela Child-Villiers, youngest daughter of the 5th Earl of Jersey, at St Pancras Church—for the second time—on the very day on which Wellington wrote. She died in 1860.

Otherwise things seemed to have quietened down in the regiment, which returned to England in 1846. At the end of that year Cardigan was promoted full Colonel. In 1848, however, there was nearly another disastrous duel. Captain the Honourable Gerard Noel quarrelled violently with his Colonel, after another parade-ground humiliation, but was directed to apologise to him by Wellington, who wrote on November 15, 1848, to Miss Burdett-Coutts:

'Lord Cardigan's affair, which gave me a great deal of trouble, has not been correctly represented in the Newspapers – as usual when the Gentlemen of the Press interfere in any Affair! The truth is that Capt. Noel was bringing the Affair originating on the Parade to be a

Private quarrel, which would have brought it to a duel. This course occasioned my Interference and the particular mode of it. . . .'

It is pleasant to record that in July 1849 Wellington was able to congratulate Cardigan in writing on the admirable discipline of the 11th, and the following year, when he inspected them for the last time before his death, the Duke in another letter to Cardigan referred to 'their excellent conduct' over the past two years.

Two years later England's greatest soldier was dead, and with him vanished an era in British military history. Another, very different series of campaigns, which would have horrified Wellington, was however about to begin— the Crimea. The 11th were now back in Ireland, and when the Cavalry Brigade was reviewed by Prince Albert in Phoenix Park in September 1853 it was commanded by Colonel the Earl of Cardigan, who was shortly to be promoted Brigadier-General.

Even with his promotion Cardigan had in no way given up command or control of the regiment, though he could leave the details to his subordinates. Mrs. Woodham-Smith describes how as he reached his fifties Cardigan, despite his long failed marriage, had achieved some sort of satisfaction, a renowned rider to hounds, a celebrated lady's man who entertained magnificently at Deene Park, but still with that fateful arrogance and temper.

> 'At last it seemed,' she writes in *The Reason Why*, 'that he had found his niche. Captain Wathen, "Black Bottle", Richard Reynolds, the Sunday flogging were forgotten. Now he had achieved his ambition; the British Army had no smarter regiment than the 11th Hussars – their precision, their operatic splendour were famous.'

And what of Black Bottle? He was continuing his Army career (he ended as a Major-General). Years later, in 1865, Major-General Desmond O'Callaghan has related how his father, who had been Surgeon with the 11th, went up to Cardigan at a regimental dinner in some trepidation and asked him on behalf of the regiment to forgive old scores and shake hands with Reynolds, who was present. Predictably Cardigan's reply was angry: 'You ought to know me better than to ask such a thing! Quite impossible!' 'But you are both old men, why nurse this miserable quarrel to the grave?' 'Will he come to me?' 'Yes, he is only too anxious to do so.' 'Bring him up, O'Callaghan,' said Cardigan. Major-General O'Callaghan tells how his father did so: 'They grasped each other's hands but neither could speak, and as dinner was announced at the same moment, the two old men, still hand in hand, walked silently into the dining-room and sat down next each other.'

Lord Cardigan

Chapter 7
The Crimea

'When the Earl of Cardigan was ordered to the East in command of a brigade of our light cavalry, from club to pothouse marvelled how he would behave. Their remembrance of him satisfied all that he had a taste for gunpowder, but they had had no experience of how he could wield a sword.' (*Our Heroes of the Crimea*, by George Ryan, Geo. Routledge, 1855, p. 47.)

'I have often been asked whether he confided to me anything particular about the Charge of the Light Brigade, but the truth is that he never seemed to attach any importance to the part he played.' (*My Recollections*, by the Countess of Cardigan and Lancastre, London, Eveleigh Nash, 1909, p. 90.)

BY 1853 Russia, Britain's ally during the Napoleonic wars, was building up her naval power and seeking access to the Mediterranean. Ever since 1844 Tsar Nicholas had been looking for some excuse to attack the large but ill-managed Ottoman Empire, and believing that the demise of 'the sick man of Europe' was near in 1853 occupied the Danubian principalities to put pressure on the Sultan, having used the rather specious pretexts of Turkish oppression of Orthodox religious groups in Constantinople, always a centre of conspiracy and unrest, and the Holy Land. Although England, not sympathetic to Napoleon III, did not want an alliance with France she was forced into it by the development of Russian sea power —the Czar had built a great naval base at Sebastopol, which openly menaced Constantinople—and despite her efforts for peace Turkey declared war on Russia in October. The British fleet had been sent to Constantinople to show the flag the previous month, to no avail, and on November 30 the Russians won a convincing naval victory against the Turks at Sinope. Russia's refusal to quit the Turkish Danubian provinces and various other events conspired to make England and France inevitably enter a war that they did not need and which could do them no good. After their declaration of war on March 28, 1854, they became committed to defend the Ottoman Empire at long range by providing the necessary land and sea forces to achieve this improbable end.

The 11th in Ireland had already received a warning order for foreign service in March, and two Service squadrons of 125 men and horses each, with another forty-five men dismounted, left Kingstown in mid-May under the command of Major Douglas, who was promoted Lieut-Colonel on June 20, the same day that Cardigan became a Major-General. In England despite or because of the ignorance of the real causes and hazards of the war, the despatch of the expeditionary force was enthusiastic and light-hearted, and a wave of patriotic fervour swept over the comfortable Victorian middle classes. Rather in the manner of the Washington socialites who drove out by carriage to watch first Bull Run, the Crimean expedition was thought of as a sort of vast and entertaining military picnic, and one or two officers' wives even sailed out with their husbands or joined them later.

The Times poked fun at the unsuitability of the uniform of the 11th for war, at 'the shortness of their jackets, the tightness of their cherry-coloured pants', which were as 'utterly unfit for war service as the garb of the female hussars in the ballet of Gustavus, which they so nearly resemble'. Predictably Cardigan gave the lie, and the main result of the lively correspondence that ensued was to make the regiment's distinctive uniform even more famous, so that when the Secretary for War, the Duke of Newcastle, wrote to Lord Raglan, who had been appointed to command the expeditionary force, on the outfitting of the troops he declared, 'I am not going to write to you about the colour and tightness of Cardigan's cherry coloured pants.' The regiment was now often nicknamed 'The Cherubims', or sometimes, more rudely, 'The Cherry Bums'. *Punch* on April 22, 1854, published a little ditty about them:

> 'Oh, Pantaloons of cherry,
> Oh, redder than the berry,
> For men to fight in things so tight
> It must be trying, very.'

At all events the 11th went to war the best mounted regiment on either side, thanks to Cardigan's purse and knowledge of horses. He was now rising fifty-seven, and about to go into action for the first time—and, more important, to command troops, the Light Brigade, in action. This consisted of the 8th and 11th Hussars, 4th and 13th Light Dragoons and 17th Lancers: the Heavy Brigade, under Brigadier-General Honourable James Scarlett, was composed of the Royal Dragoons, the Scots Greys, 4th and 5th Dragoon Guards, and the Inniskilling Dragoons. These made up the Cavalry Division, to which were attached Maude's Troop R.H.A., a battery of Field Artillery, the 93rd Highlanders under Sir Colin Campbell, 1,200 British marines under Colonel Hurdle, and 3,000 Turks under Rustem Pasha. The whole was commanded by Lord Lucan, Cardigan's brother-in-law.

The glories of Waterloo and the Peninsula were now but distant drum-beats; for years England had been at peace, and only those who had served in India had any appreciation of active service, let alone war conditions. Lucan fifty-four, Scarlett fifty-five, Cardigan fifty-six—these were men with only a few years of middle age left, and only Lucan had seen a shot fired, and that over twenty-five years ago, curiously enough in much the same area as he now found himself in. The C-in-C, Lord Raglan, who had been Wellington's Military Secretary, and an admirable one, had lost an arm at Waterloo and neither his bravery nor intelligence were doubted. Yet all his experience had been as a staff officer and he had no background of commanding troops, far less armies, in action. Worst of all, Raglan, who was a family friend of the Cardigans, had ill-advisedly let Cardigan believe that though undoubtedly junior in rank and position to Lucan he could more or less operate on his own with the Light Brigade. This suited Cardigan excellently, since the brothers-in-law could not abide one another, but was in no way dreamed of by Lucan, who was determined to make his military reputation. Today it seems utterly incredible that Whitehall could have countenanced both appointments: as W. H. Russell, the distinguished correspondent of *The Times*, wrote later:

> 'Lord Lucan was a hard man to get on with but the moment the Government of the day made the monstrous choice of his brother-in-law, Lord Cardigan, as the Brigadier of the Light Brigade in the Light Cavalry Division, knowing well the relations between the two officers and the nature of the two men, they became responsible for disaster; they were guilty of treason to the Army—neither more nor less.'

The 11th after a long and uncomfortable voyage in sailing ships through the Mediterranean, the Aegean, the Dardanelles and the Bosporus, in which the horses suffered considerably, landed at the Bulgarian port of Varna towards the end of June. One of those in the barque *Paramatta*, which conveyed sixty soldiers and horses all the way from Kingstown pier (where they were seen off by the Irish with 'whisky galore') was Private W. H. Pennington, who later became a noted Shakespearian actor (and friend of Henry Irving) whom Gladstone admired. Pennington had enlisted at Dublin in 1853, and soon came

Army Museums Ogilby Trust
A wounded veteran of the Crimea.

to love his regiment. His comments about the relationship that existed between the officers and the men are of extreme interest, in that they go far towards discounting the popular picture of military life at the lower levels under Cardigan and the 'Haw Haw' ineptitude of lisping officers depicted in some films. 'I think all of us voted our officers "jolly good fellows"; few soldiers ever had a happier or less harassing time,' he told Mrs. Tom Kelly, and again:

> 'I should hardly think that regimental records could furnish a stronger instance of good feeling than that which existed between these gallant young officers, and the rank and file under their command. I do not remember any instance of punishment; indeed I fail to remember an ungracious word.'

In his own book, *Sea, Camp and Stage* (1906) Pennington records,

> 'our officers were such good fellows. Captain Inglis, Lieutenant Trevelyn (subsequently colonel of the 7th Hussars, but now, I regret to say, deceased), Sir Roger Palmer, Bart. (not in possession of the title then, but now retired lieutenant-general), were all officers incapable of a harsh exercise of power.'

Pennington's opinion of Cardigan, coming from a ranker, is of more interest than that of most of the officer or newspaper memoirists of the Crimea:

> '. . . the overbearing manner of the unfortunate Brigadier rendered him unpopular with every officer under his command. By the rank and file of the 11th Hussars he was known as "Jim, the Bear", and they with a somewhat extravagant opinion of his gifts as a cavalry officer regarded him as the Murat of the British Army; but by what standard of judgment they arrived at such a measure of his capacity I have failed to learn, for he never gave any evidence of those attributes of prescience or inspiration which at an important crisis mark a great military leader.'

Cardigan had already arrived by steamer and yacht at Varna (having been entertained by Napoleon III at the Tuileries en route), suffering rather badly from bronchitis, though not badly enough to prevent his setting up his headquarters in a pleasant house across the bay at Devna, where most of the Light Brigade and some of the Heavy Brigade and numbers of Turkish cavalry had now congregated. This enabled him to ignore with pleasure Lord Lucan, still tormenting his staff with a torrent of orders near Scutari, and to deal directly with Raglan, who had also arrived at Varna. It was insufferably hot and dirty in Bulgaria, the sanitation was more or less non-existent, and soon the troops were suffering badly from epidemics of cholera and dysentery. The Rev. H. P. Wright (*Recollections of a Crimean Chaplain*) recalled: 'The mortality at Varna was quite appalling. I buried twice, and sometimes thrice, daily; and my average, including deaths in the small encampments, was nine.' Many horses had been injured or fell sick in their unaccustomed surroundings and lacked suitable food. Major Forrest, now in the 4th Dragoon Guards, saw which way the wind was blowing, for there was a series of acrimonious letters between Cardigan and Lucan, the latter rightly assuming that Cardigan was determined to ignore him if not supersede him. Forrest thought a row between the two inevitable, and while appreciating Lucan's cleverness felt no confidence in him, for one thing because 'he has been so long on the shelf he does not even know the words of command'. But when Cardigan, intent on operating in a self-contained command, rudely told Lucan he intended to get Raglan's authority to do so, Lucan formally complained to the C-in-C. As Russell thought at the time, Raglan should have removed one if not both of the cavalry commanders forthwith. Instead he did nothing except attempt conciliation through his Adjutant-General, and, when the Russians were driven out of the border town of Silistria by a Turkish force led by British Indian Army officers, ordered Cardigan to make a reconnaissance into the Dobrudscha to find out how far the enemy was retreating across the Danube.

On June 25 Cardigan took two squadrons, of the 8th Hussars and 13th Light Dragoons, to Karasu and Rassova, and patrolled the Dobrudscha. He reached the Danube on June 29 to find the Russians gone. His small force rode along the banks of the river to Silistria, and returned to Devna on July 10 by way of Shumla, Yenibazaar and Pravadi. Cardigan himself described the expedition:

> 'We travelled over the country which I may call a perfectly wild desert, for a distance of 300 miles. My orders were to proceed 130 miles as far as Trajan's Wall, on the confines of the Dobrudscha. We did so, and marched 120 miles without ever seeing a human being. There was not a single house in a state of repair or that was inhabited all along this route, nor was there an animal to be seen except those that exist in the wildest regions.'

Across the Danube, however, the Russian General Luders had often observed the mounted body with interest through his telescope, and though it came within range of his guns he forbore to open fire. The reconnaissance cost the Light Brigade dear in horse casualties—nearly 100 out of 280 from lack of water and forage and from overwork in the heat—and was nicknamed 'the sore-back reconnaissance'. But Raglan was pleased by what it seemed to have found out.

Back at Devna Cardigan once more concentrated on spit and polish. Captain Cresswell, who had brought his wife with him, and was one of those shortly to die from cholera, wrote to a friend describing how 'the Major-General amused us by giving us regulation Phoenix Park Field days —such a bore he is—comes round stables just as if he was Colonel instead of Major-General, and he makes us all go to evening stables.' Lucan meantime was deluging his brother-in-law with sheaves of unnecessary returns and directions; those of which he did not ignore Cardigan returned unsigned or incomplete. Then at the end of July he scored off Lucan by suddenly moving the Light Brigade out of Devna without orders, settling at Yenibazaar some sixteen miles from Shumla. Cresswell called it 'a splendid place', and as at Devna the surroundings were picturesque, but once again in Pennington's words 'fever, dysentery and cholera were our grim companions, and carried off their victims freely every day. Some battalions lost nearly a third of their strength, and in the Vale of Aladyn, as beautiful a spot as could be found on earth, the Guards were suffering more than any corps.'

Ignorance of what they were supposed to be doing ('No one knows what we are here for, or what we are going to do. . . . As to fighting, it is the last thing we think of . . .' wrote Cresswell) added to sickness sapped some of the men's morale, but Cresswell, in the last letter of his life on August 14 from Yenibazaar mentions that the 11th had kept very healthy, with only twenty men sick, compared to sixty of seventy in the 8th and 17th. With a sad irony he tells his friend back home that they are suffering more from 'fever of country' than from cholera, and that there is no glory to be got out here, only discomfort'. As always in periods of inactivity there were grumbles and rumours:

> 'We hear that Sebastopol is to *fall* and that the only Cavalry going is the 11th P.A.O.H. but I doubt it. We go on here in the same indolent way. Smoke and sleep all our spare time. Only fancy? The Major General has sent *every* officer of the Brigade, to Morning Stables, and an Officer per troop to evening stables so you never get away.'

Raglan's force and the French under Marshal St Arnaud had still to strike a blow at the Russians after three months, and although the Russian retreat across the Danube had given an opportunity for peaceful negotiation, public opinion in England demanded some signal success from the large and expensive expeditionary force. Mistakenly giving in to this pressure, knowing almost nothing of the dispositions and strength of the Russians (except that the latter was bound to be great) or of the defences of the great naval base at Sebastopol (except that they were bound to be formidable), with no knowledge of the coastline, with an army decimated by disease and lacking in equipment and supplies, Raglan, with the French consenting, decided to attack in the Crimea. The decision would have appalled his former chief.

Thus on August 24 the chaotic embarkation into overcrowded ships at Varna began; 24,000 French infantry with 70 guns, 6,000 Turkish infantry, 22,000 British infantry, with 1,100 cavalry and sixty guns, were crammed into the transports. The French, whose organisation was infinitely superior both at sea and on land, left first, the English on September 7, and on the 10th both fleets rendezvoused forty miles west of Cape Tarkan. Some sort of sea reconnaissance had been attempted by Raglan and General Canrobert steaming down the Crimean coast, and after various arguments an extended beach six miles north of the Bulganak river was chosen for disembarkation, two areas two miles apart being allotted the two armies. The unfortunate name of the landing ground was Calamita Bay. Raglan in the steamer *Caradoc* and some of the French staff had already taken a close look at the Russian troops at Sebastopol, whose officers could be seen in turn observing them through field-glasses. The British had saluted gravely and the Russians had returned the compliment.

On September 14 the disembarkation started early, and for most of the day it went well. In the evening torrential rain and a rough sea made things most unpleasant, and many men were taken ill during the night for they had no shelter and little food. Cholera was again rife, and there was a shortage of fresh water. The incompetence and inexperience of Raglan's staff was to blame for many lapses in organisation, and E. H. Nolan records:

> 'The condition in which the British Army found itself when the embarkation was completed, and the march into the enemy's country begun, was pitiable in the extreme. There were no tents, the officers had no horses, the men were without their knapsacks, the medical men without proper supplies of medicine, bandages, or other means of adequate surgical or medical treatment. Destitution and disorder reigned in the British Camp, so far as any arrangements were concerned that were not directly military; while the French exhibited an organisation nearly perfect.'

Curiously enough the Russians did not try to interfere with the disembarkation, or to attack once Raglan and St Arnaud started the dry, dusty march towards Sebastopol. In numbers of cavalry they were vastly superior, for the French had none, the Turks looked like 'beggars on horseback', according to Nolan, and would have been better off on London cab-horses, and for the moment Raglan had left behind the Heavy Brigade at Varna. This left about 1,100 men of the Light Brigade, of whom the 11th had the best horses; when on September 18 one of its troops (Cresswell's) ran into seven times its number of Cossacks on reconnaissance, it made rings round them, and Nolan commented, 'had all our Light Cavalry been mounted like the 11th Hussars, their service would have been more useful. . . .'

Mostly the Cossacks were content to watch at this stage, and to burn villages in the line of advance. On September 19 the Allied force moved cumbersomely forward. It was excessively hot, and many of the infantry fell out from exhaustion and thirst; when the Bulganak stream, some twenty miles from Sebastopol, was reached whole divisions flung themselves gratefully towards it. Cardigan now pushed out squadrons of the 8th and 11th Hussars and of the 13th Light Dragoons to investigate the Bulganak valley; they soon saw Cossack lances flashing in the sun on the hills beyond. Both sides sent out skirmishers, but Cardigan, as Lieutenant George Peart of the 20th Regiment relates, soon discovered that the Russians had immense columns in reserve—there were in fact some 6,000 men of the 17th Division, a brigade of cavalry and artillery—and, quite rightly, 'thought that our cavalry would be blown in ascending such a hill and very likely surrounded and cut to pieces by a force three times their strength'. This was not the reaction of a dashing fool. Cardigan therefore called in his skirmishers and ordered a slow retirement, which Colonel Douglas sensibly conducted at a walk rather than the usual trot. The last thing wanted at the moment was for the British to get involved in a battle against greatly superior odds in the wrong place, so Raglan ordered up the 2nd and Light Divisions and some nine-pounders to support Maude's six-pounders. Cardigan's squadrons were fired on with little effect, and there was some exchange of artillery fire, in which the British came off best, but all through the withdrawal of the cavalry there were two squadrons facing the Cossacks. Most of the enemy losses were inflicted by the carbines of the 11th. E. H. Nolan, *The History of the War against Russia*, says:

> 'It was afterwards ascertained that twenty-five men and thirty-five horses belonging to the enemy were hit. The carbines of the 11th Hussars principally inflicted this loss, for being so well mounted, they rode in small detachments, very near to the Cossacks, and fired, retiring with impunity.'

The Russian artillery fire was mostly inaccurate, but some of the young regimental soldiers covering the R.H.A. became nervous and bowed their heads down to their motionless horses' manes, Major Yates Peel cried 'What the hell are you bobbing your heads at?' The first casualty in the British force was Private Williamson of the 11th, whose foot complete with stirrup-iron was taken away by a round shot. Sir Roger Palmer, his Troop Commander, said, 'He rode out of the ranks, his leg shot off and hanging by his overall. Coming up to me he said, quite calmly, "I am hit, may I fall out?"' Williamson died aboard ship on his way to hospital at Scutari.

The Bulganak position, where the British bivouacked in order of battle that night, was not a good one, and as a gap had occurred during the day's march between Raglan and the French, a determined flank attack by the Russian commander, Prince Mentschikoff, might well have been successful. But Mentschikoff did nothing. On September 20 the Allied armies linked up again, and pushed on to the Alma stream, behind which on the hills and ridges a large Russian force was spread over four miles, its left flank resting on a high wall of rock that led to the sea.

Raglan was determined not to risk his precious cavalry, mindful of near disastrous charges at Waterloo and in the Peninsula. Like his great master he was desperately short of them compared with the Russians, just as Wellington had usually been short compared with the French. His task was not made easier by the continuing friction between Lucan and Cardigan; the former, having at last caught up with the action, was determined to allow the latter no rope at all, and constantly interfered with his arrangements and dispositions.

Lucan had already interfered at the Bulganek, though Raglan's Alma despatch was to state that Cardigan had 'exhibited the utmost spirit and coolness, and kept his Brigade under perfect command'. Nevertheless, it had been galling in the extreme for the cavalry not to be allowed to engage the enemy, but to have to retreat under the jeers of the Cossacks, and henceforth, unjustly, Lucan was to be known as 'Lord Look-on', since it was thought by the troops that he had been to blame for holding back. Cardigan, in his turn, was shortly to become 'The noble Yachtsman', from his habit of sleeping and eating aboard his yacht, the *Dryad*, whenever possible. Already, however, Cardigan had been slightly wounded in the right leg by a Russian lance during one of the skirmishes, and if one of his subalterns had not struck it down might well have been killed. He had displayed coolness when the 11th had been shot at in bivouac by an infantry regiment which mistook them for enemy, and was to do so again on the Alma heights when a rambling cannon-shot leapt over the croup of his horse.

The 11th led the advance from the Bulganek to the Alma, and drew the first fire from the village of Bourliouk in front of the river. Despite the excellence of the natural defensive position, by a series of errors including the premature withdrawal of their artillery, and primarily through the extraordinary and bloody courage of all the British and most of the French infantry, Alma turned into a decisive Russian defeat. But for the British cavalry it was a day of agonising frustration. For once Lucan and Cardigan, powerless to move without orders, were in agreement, and even when the Russian infantry broke, Raglan, over-anxious not to risk his cavalry but to keep them 'in a bandbox', would not loose them to turn defeat into a rout because there was not enough artillery support and he feared a trap. All they were permitted to do across the river was to take a few prisoners and stragglers, but Raglan through Major-General Estcourt had categorically ordered Lucan 'The cavalry are not to attack.' Worse still, when the field was clearly seen won, the French declined to join a pursuit, and Raglan dared not go on alone. Thus was wasted a chance of complete victory, and the Russians were allowed to stream away to Sebastopol.

Of all the frustrated and impatient cavalrymen, none was

more furious than Captain Lewis Nolan of the 15th Hussars, an outstanding horseman and officer who was chief A.D.C. to the Quartermaster-General, Airey, and thus in close contact with Raglan. After Alma, Nolan fumed to Russell of *The Times*,

> 'There were one thousand British cavalry, looking on at a beaten army retreating—guns, standards, colours and all—with a wretched horde of Cossacks and cowards who had never struck a blow, ready to turn tail at the first trumpet, within ten minutes' gallop of them. It is enough to drive one mad! It is too disgraceful, too infamous....'

And the impetuous Nolan blamed Lucan and Cardigan; but it was not their fault, and Lucan had already been moved to request Raglan to let him act on his own responsibility as commander of the cavalry lest opportunities for action be lost—a request which Raglan did not see fit to answer. What Lucan and Cardigan were most to blame for was their constant guerilla warfare against each other at the expense of the troops they commanded, 'like a pair of scissors who go snip and snip and snip without doing each other any harm, but God help the poor devil who gets between them', as Lord George Paget, commanding the 4th Light Dragoons, wrote home.

Raglan's despatch on the Alma was less than honest when he commented that the nature of the ground did not permit the use of cavalry. Russell disputed this later, stating 'The ground *was made for cavalry* from the Alma to the Katcha!' The war correspondent described Nolan as 'impetuous, vehement, restless', and professed astonishment at the angry way in which Nolan spoke of Lucan and Cardigan, especially of the former, though after 'the soreback reconnaissance' to the Danube he was full of invective against Cardigan as well.

Although there had been fairly heavy casualties, the obvious thing for the Allies now was to press on immediately and attack Sebastopol from the north; the Russians were in a state of nerves, their defences ill-organised, and the British and French fleets could have helped by bombardment. Raglan, however, though inclined to this course would not jolt the ailing St Arnaud, the French inclined to an attack from the south and were supported in their view by the aged British General Burgoyne—an illegitimate son of the famous 'Gentleman Johnny' who surrendered to the Americans at Saratoga and who, as a young officer, had been in the 11th—and in the end this opinion prevailed, disastrously. In fact had they known it the Allies could almost have walked into Sebastopol from the north, despite its reputedly powerful fortifications on which the great military engineer Todleben was desperately working—for the good reason, as Todleben himself explained, that the defences had not been prepared and were not yet properly manned.

On September 21 and 22 the regiment had patrolled beyond the Katcha river, and the next two days were spent in a full-scale cavalry reconnaissance to the village of Duvankoi and the river Belbec, by which the 11th bivouacked on the 24; with the 13th Light Dragoons it had reached the causeway crossing the marsh below the heights of Inkermann on the river Tchernaya. Opposition was light and disorganised: now above all was the time to attack Sebastopol a few miles away to the north. St Arnaud, already a sick man (shortly to die of cholera, which was still prevalent) exaggerated the difficulties that Fort Constantine on the north of the city would present; Raglan, unable or unwilling to shake the obstinacy of a dying man, mistakenly abandoned his faith in a northern attack and proposed that the allies should make a long flank march through unknown woods to the east of the city and then begin the siege from the south. St Arnaud agreed. The only conceivable advantage of this time-wasting plan was to offer a firm sea base at Balaclava.

The flank march started inauspiciously when a staff officer guiding the cavalry lost his way and the whole of Lucan's force except the horse artillery took the wrong turning in thick forest. Raglan, Airey and his staff coming up forward took the right route, towards the ruins of Mackenzie's Farm, named after a Scottish admiral, east of the city near the high road to Bakshiserai; Airey, in front, nearly ran straight into the Russian rearguard, and Raglan, joining him, gazed calmly for some minutes at the surprised Russians, who, if they had had their wits about them, could have captured the British C-in-C there and then. With the main body of cavalry still trying to get back on the right route the horse gunners saved the day by quickly blasting the Russians below them at point-blank range—one regiment of infantry on the road was only thirty yards away—and Lucan's cavalry finally arrived to complete the Russians' disorder. Raglan, however, called them off any proper pursuit. Peard relates:

> 'The Cavalry, consisting of the 8th and 11th Hussars, were quickly sent out, and the guns brought to bear on the astonished convoy. The Rifles [of the 1st Division], in skirmishing order, kept up a heavy fire on them, and the Cavalry making a charge, they were completely routed, leaving behind them an immense quantity of baggage for two miles in the direction of their flight. This was lawful plunder, and the soldiers were allowed to break open the carts and help themselves, under the superintendence of an officer, to prevent squabbling. Wearing apparel of all descriptions was found in great abundance, together with dressing-cases, jewellery, ornaments, and wine, and I heard that some specie was also taken. Fine large cloaks, lined with fur, and some hussars' jackets, covered with lace, were also seized and sold by the soldiers. Menschikoff's carriage likewise was taken. . . . This put our men in the highest possible spirits, and enlivened the long, dusty march to the Black River, where they halted and bivouacked for the night.'

By September 26 the British and French were encamped on the plateau above Balaclava, the British facing Sebastopol from the south-east, the French from the south-west with convenient and easy supply bays at Kamiesh and Kazatch. The British base at Balaclava, however, soon became overcrowded and was five to seven miles from the troops; supplies were short and late, and lack of transport animals meant that cavalry horses had to be used for unsuitable jobs. And yet the weather was lovely, the countryside beautiful, and for the moment sickness not as bad as it had been, so that the soldiers' spirits were high. Even now it was not too late for a bold frontal attack on Sebastopol, which Sir George Cathcart, G.O.C. the 4th Division, strenuously advocated. Cathcart was right, but because he was not on good terms with Raglan and because the French, even under their new commander, Canrobert, were not ready to make an assault, it was decided to lay formal siege to the city, and digging and the ponderous unloading and siting of siege guns began. Meantime in Sebastopol Todleben and Admiral Korniloff were making furious and successful efforts to strengthen their defences and supplement their men, and the one priceless commodity they needed was now presented freely by Raglan and Canrobert—time.

Much of the digging in the British sector was ineffective because the ground was too rocky, their engineers ill-equipped, but the French spiritedly got some guns within a thousand yards of the city's central bastion. But the long-awaited bombardment which opened at last on October 17 was disappointing and unco-ordinated with the guns of the combined fleets offshore. The French bore the brunt of the Russian counter-battery fire, and suffered heavy casualties, and several British ships were badly damaged without inflicting substantial damage themselves on the enemy forts. Yet by evening when both the Malakoff and the Redan defences were seemingly nearly shattered a combined assault would still probably have won the day. None was made, because the French were disheartened and unready and there was no supreme commander to give the order. For a week the bombardment continued, but as fast as Russian defences were knocked down they were rebuilt and strengthened.

As the days dragged by cold and rain ominously began, and cholera and diarrhoea sapped the strength of thousands. Still no assault was made, for the French were disheartened by severe Russian shelling of which they endured the worst, and Raglan began to worry about a counter-attack on his exposed right flank and Balaclava. Morale slipped, and the cavalry, kept in the saddle day after day but never allowed to chase the always observing Cossacks on the Tchernaya and in neighbouring villages, were becoming impatient and resentful, the more so as horse casualties were serious. Angry words passed between Nolan and Lucan when the latter, still hamstrung by Raglan's orders, refused to commit his cavalry against a large Russian force manoeuvring in the Balaclava plain on October 7, and Cardigan, who had been sick, returned to duty to criticise not only his brother-in-law, as usual, but, more justly, the conduct of the whole siege. Now resident, with Raglan's permission, by night on his yacht, and constantly accompanied by his civilian friend Hubert de Burgh, Cardigan averred that 'I have never in my life seen a siege conducted on such principles', in no way inhibited by his own inexperience, and one day when the pair encountered *The Times* correspondent near Balaclava William Russell related Cardigan's comments:

> 'Haw! Haw! Mr. William Russell! What are they doing? What was the firing for last night? And this morning?' I confessed ignorance. 'You hear, Squire? This Mr. William Russell knows nothing of the reason of that firing! I dare say no one does! Good morning!'

By now the Heavy Brigade, under Brigadier Scarlett, consisting of the Royal Dragoons, the Scots Greys, the 4th and 5th Dragoon Guards, and the Inniskilling Dragoons, had arrived. Lucan at last had his full Division, even if Raglan detached the 11th and the 17th Lancers to form a separate camp on the heights beyond Balaclava under Cardigan's direct command—though usually its administration fell to his second-in-command, Lord George Paget of the 4th Light Dragoons, and his Brigade Major, Lieut-Colonel Mayow, with both of whom he had several 'dust-ups'. Paget liked and supported Lucan, but knew how to

Army Museums Ogilby Trust

A rare photograph of General Lord George Paget taken during the Crimean War.

deal with Cardigan ('He is easily managed, with calmness and firmness, and when one is in the right—which it is not difficult to be with him') and avoided quarrelling with him. Even the level-headed Paget, however, could not refrain from describing his two senior commanders as 'these two spoilt children' whom it was found desirable to separate because they were 'again hard at it'. Captain Robert Portal, a fellow officer, thought that Cardigan 'has as much brains as my boot. He is only to be equalled in want of intellect by his relation the Earl of Lucan. Without mincing matters, two such fools could hardly be picked out of the British Army. And they take command. But they are Earls!' Forrest, the former 11th Hussar, wrote: 'We all agree that two greater muffs than Lucan and Cardigan could not be. We call Lucan the cautious ass and Cardigan the dangerous ass.'

★ ★ ★ ★ ★ ★

The Russians outside Sebastopol had not been idle. By October 24 a mixed force of 25,000 men under General Liprandi had collected round Tchogoun, ready to attack the British at Balaclava, which, two and a half miles away from the main siege position and only lightly garrisoned, was still essential as the only supply point and harbour. Only one road, the Worontzoff road, ran from Sebastopol, seven or eight miles away, towards Balaclava, and it crossed the Plain of Balaclava on a ridge of land known as the Causeway heights, beyond which to the north lay the North Valley and the Fedioukine Hills, while to the south was the South Valley and some isolated heights including Kadikoi. Each valley was shut in at both ends by steep hills. The Causeway Heights, providing the only line of communication between the British base and the forward camp, led also to the Upland and the Sapouné Ridge whence the allies besieged Sebastopol. It was thus vital. But it was defended only by six artillery redoubts each with a 12-

pounder gun, manned by upwards of 1,000 Turks and Tunisians—the latter 'unaccustomed to war'—with a few British N.C.O.s to encourage them, while Balaclava itself, some three miles away was held only by 1,000 Marines and men recovering from sickness. In between on some high ground near Kadikoi was the only first-class fighting unit, 550 men of the 93rd Highlanders, and two more Turkish battalions. Despite a warning from a Turkish spy on the night of October 24 that the Russians were about to attack Raglan ignored this accurate intelligence (it was by no means the first warning). The cavalry camps were to the left front of the Highlanders in the valley below the Upland.

At 5 a.m. on October 25, about the time when the cavalry were turning out, the Russians began their advance, seized the village of Kamara, and installing their heavy artillery on the Fedioukine Heights began to attack the three eastern Turkish redoubts with greatly superior numbers. Simultaneously their main body of cavalry poured into the North Valley. The Turks on Canrobert's Hill, the site of No. 1 Redoubt, put up a sturdy defence, but with one battalion against tenfold odds they were driven from their position with 170 men killed; so too was Maude's Troop, R.H.A. which with some Scots Greys had come up to help them. Disheartened by their countrymen's defeat the Turks and Tunisians in the next three Redoubts put up poor resistance, and some, losing their nerve, made off *ventre à terre* for Balaclava; others, according to Paget, crawled slowly by laden with blankets and kits. Paget thought their bad conduct much exaggerated, and claimed with reason that the best troops in the world should not have been left unsupported in such positions. Russell, however, recorded that the Turks 'fled with an agility quite at variance with common-place notions of Oriental deportment on the battle-field'. Now was the time for Liprandi to descend on Balaclava, skirting the 93rd if necessary, for as yet there was no sign of either the 1st or 4th British Divisions which Raglan had ordered to help when he heard of the attack. Indeed over an hour and a half was to pass before either could make its awkward way from the Sapouné Ridge. But although more and more Russians deployed across the Causeway Heights and on to the Worontzoff road, Liprandi did not turn towards Balaclava itself in any haste. The main body of over 3,000 Russian cavalry advanced up the North Valley with thirty-two guns, and then checked when a battery from the Sapouné Ridge fired at them. Then several squadrons turned left to reconnoitre the gorge of Kadikoi, between which and Balaclava lay only Sir Colin Campbell's small force of Highlanders. The main Russian mass watched these squadrons cross the Causeway Heights towards the South Valley.

Raglan had ordered Lucan not to commit his cavalry until the infantry divisions arrived—which they had not yet done—but a part of Scarlett's Heavy Brigade had already been ordered up to support—too late—the Turks in their redoubts. The Light Brigade was formed up complete less than a mile away on the edge of the uplands to the left of Redoubt No. 6 in the path of the Russian advance. At first the Russians could see neither the Highlanders nor three squadrons of Scarlett's (two from the Scots Greys, one from the Inniskillings), but when the leading cavalry crossed the Worontzoff road they perceived the former and attacked. The gallant 'thin red line' of the 93rd beat them back more than once with well-controlled musket-fire, and the Russians retreated. The main body of Russian cavalry now veered left also and crossed the Causeway Heights at an oblique angle to the Light Brigade, neither seeing the other. But suddenly some way below the Russians appeared Scarlett's few squadrons making their way in column of threes across broken country. The Russians had the advantage of the ground, but for some reason after beginning their descent did not charge but came to a halt, while two detachments were sent out to outflank Scarlett. He seized his chance, wheeled his 300 dragoons into line, having despatched a frantic message for the rest of the brigade to join him, and charged.

Although greatly outnumbered, the Heavy Brigade shattered the Russian squadrons, and for a time virtually disappeared in a seething mass of fighting horsemen. The opportune reinforcement first by the Royals and then by the two Dragoon Guards regiments from the flank turned the trick and the Russians first gave way and finally turned and galloped back over the Causeway Heights. The British casualties were light, about 110 killed and wounded, the Russians, perhaps 200—but the morale effect was immense. The 11th, of course, although less than half a mile away on higher ground, had taken no part in the action, for Cardigan, although seething with impatience, had received no orders from Lucan or Raglan freeing him from the explicit brief to stay where he was. One man from the 11th, however, Pennington relates, uniquely participated in both charges this famous day. Private Hope, a Welshman, was a prisoner in the guard tent in the cavalry lines that morning, having committed some slight offence; when he found himself deserted he 'mounted a horse belonging to the Scots Greys, and took part with that regiment in the heavy charge against the Russian cavalry. Returning unscathed, he then made for his own regiment, riding with the Eleventh into the "Valley of Death", and bearing himself like a knight errant at the Russian guns.' Hope thus won immortality.

Cardigan was much criticised, at the time and later (especially by Kinglake), for not committing the Light Brigade from the flank and by pursuit turning the Russian retreat into a rout. Indeed Captain Morris, commanding the 17th Lancers, strongly urged pursuit upon Cardigan, who peremptorily refused. But Cardigan considered, rightly or wrongly, that his orders were rigid, permitting of no opportunism (although Lucan afterwards denied this), and

Cardigan's cast of mind was also rigid. Thus a golden opportunity was lost.

It was still early in the morning, about normal breakfast-time, though few of the troops had had any breakfast or indeed had watered their horses. The Light Brigade had by now been dismounted nearly an hour and a half, and were very much at ease. Officers, flasks in hand, were munching whatever they had, or smoking, and the men, as they stood by their horses, were chatting as if they were off duty. The 4th Light Dragoons, better organised than most, enjoyed biscuits, hard-boiled eggs, rum and cigarettes.

To Raglan, accompanied as usual by his favourite mounted orderly, Private Ashe of the 11th, and his staff up on the Sapouné Ridge high above, the whole spectacle of the green valleys and soft plains far below, and further still the placid waters beyond Balaclava glinting in the sun, the gorgeous uniforms and the ant-like distant infantry, all this must have seemed almost like a staged tattoo rather than bloody battle. The immediate danger was past, but Raglan, Russell noted, was not his usual composed self. Worried by the dawdling progress towards the road of Cathcart's and the Duke of Cambridge's infantry divisions, his artillery firing at too great a range to be effective, Raglan saw that it was high time to recapture the redoubts that the Russians still held and the Causeway Heights. The enemy had artillery and infantry in strength on the Fedioukine Ridge north of the North Valley, but the mass of their cavalry and guns had now returned to the east end of the North Valley. But it seemed probable that they might try to move the captured cannon from some of the redoubts. Raglan now sent a new order to Lucan:

'Cavalry to advance and take advantage of any opportunity to recover the heights. They will be supported by the Infantry which have been ordered to advance on two fronts.' Lucan's copy of this order omitted the word 'to' before 'advance' and had a full stop after 'ordered'. Pardonably, Lucan thought that he should wait until the infantry arrived before he attacked the redoubts, and three-quarters of an hour was wasted, but not before the Russians began to move the captured guns. If they were successful that would look bad. Raglan thus despatched a further urgent order by Nolan to Lucan, reading:

> 'Lord Raglan wishes the Cavalry to advance rapidly to the front and try to prevent the enemy carrying away the guns. Troop of Horse Artillery may accompany. French Cavalry is on your left. Immediate.'

This was signed by Airey, the Quartermaster-General. It turned out to be a fatal order. Raglan cannot be held guiltless, even if Nolan's arrogant and impulsive delivery of it was equally responsible for disaster. From where Raglan sat the picture was crystal clear: but things were

Army Museums Ogilby Trust

'Lord Cardigan receiving Lord Raglan's order from Captain Nolan at Balaclava.' The picture, no more than an artist's impression, was painted in 1889.

very different hundreds of feet lower down, where Lucan and his staff in the low ground at the western end of the North Valley, between his two cavalry brigades, were cut off by the terrain from the Russians and could see in fact neither guns nor enemy. The Russians, moreover, were deployed in several areas, and there was indeed no one single front.

No wonder that Lucan, and in turn Cardigan, were perplexed and dismayed. Lucan slowly read the order which made no sense to him, and then objected that it was both useless and dangerous. Nolan, who loathed Lucan and despised his abilities, shouted at him 'Lord Raglan's orders are that the Cavalry should attack immediately!' Lucan, flustered in turn, replied sharply: 'Attack, sir! Attack what? What guns, sir?' Nolan now committed the unforgivable error, flinging out his arm and pointing in disrespectful rage: 'There, my Lord, is your enemy! There are your guns!' But Nolan's arm was pointing, not at the redoubts and the Causeway Heights, but up the long North Valley to the end of which the Russian cavalry had retired and where they now had formed up with their artillery batteries in front of them. Nolan's whole conduct, and his remarks to Lucan and then to Cardigan, were in the memory of the Duke of Cambridge so 'offensive' that although both doubted Raglan's intentions and foresaw what might happen they saw no choice but to carry out the garbled order. Lucan with icy politeness had trotted over to Cardigan at the head of the Light Brigade and passed on the exact wording. With equal politeness Cardigan dipped his sword in salute, and then remonstrated strongly in his husky penetrating voice: 'Certainly, sir, but allow me to point out to you that the Russians have a battery in the valley to our front, and batteries and riflemen on each flank.' Then he added, according to Corporal J. W. Kilvert of the 11th, 'There must be some mistake. I shall never be able to bring a single man back.' 'I know it,' said Lucan, 'but Lord Raglan will have it. We have no choice but to obey.' Cardigan saluted: 'Very good, sir.'

He then rode off to the head of his brigade and formed it up as if on a parade ground in two lines, the 13th Light Dragoons on the right in front, the 17th Lancers in the centre front, the 11th Hussars on the left slightly to the rear; the second line consisted of the 4th Light Dragoons and the 8th Hussars. None of the regiments was anything like up to strength, because of sickness and casualties, the 17th with 145 men being the strongest; with staff and officers the total mounted soldiers in the charge is given as 673, but the regimental parade states numbered only 607. Just before the charge Lucan, without consulting Cardigan, ordered Colonel Douglas to withdraw the 11th into the second line, the 4th and 8th thus becoming the third. Two lengths in front of his own small staff, and eight ahead of the first line, Cardigan sat for a moment bolt upright on his favourite thoroughbred chestnut, Ronald, whose two white legs were clearly seen by the watchers on the Sapouné heights. Muttering half to himself, 'Here goes the last of the Brudenells,' Cardigan, with no trumpet sounding, gave his first and almost last orders of the charge; they were a few rapped out words: 'The Light Brigade will advance—Walk—March—Trot!'

One of the last arrivals to join the brigade was the butcher of the 17th Lancers, Jack Veigh, still in shirt sleeves, who had been employed in more peaceful duties in the lines. Some of the men in the 8th Hussars were smoking pipes, though their Colonel told them to extinguish them; Paget, leading the 4th Light Dragoons, had a cheroot between his teeth, and kept it there until he reached the Russian guns. It was, according to Russell's watch, ten minutes past eleven in the morning as the Light Brigade moved off. All along the beautiful ominous valley could be heard nothing but the jangle of equipment, the hooves and snorting of the chargers, as the lines of horsemen moved, slowly at first, then in a fast trot. Nolan, who had received permission to ride with his friend Morris, suddenly realised after a few moments the ghastly mistake that was being made, through his own chief fault, and putting spurs to his horse rode diagonally across Cardigan's front waving his sword in the direction of the Causeway Heights. This infuriated Cardigan, who did not understand what Nolan was trying to do—and it was all too late, for the next moment the silence was broken as the Russians opened a fierce fire from front and flanks, and one of the first shells mortally wounded Nolan, whose dying and then lifeless body was carried back through the 13th Light Dragoons still astride his mount.

Erect and splendid, Cardigan rode on unmoved, always checking the brigade from going too fast, almost as if on parade. Not until the Russian guns were nearly reached were any horses galloped, and at one time when a squadron of the 17th Lancers began to force the pace and its leader, Captain White, came up level with him Cardigan gently laid his sword sideways across the officer's breast and ordered: 'Steady the 17th!' Kilvert recalled, 'In going down the valley Lord Cardigan kept his Brigade well in hand and checked the pace several times, and I can state positively that we went down the valley at a steady canter.'

And what were the feelings of the men in the charge? The Russian round shot and shell fire, supplemented by Minié balls from the infantry's rifles, was murderous, and came not only from the end of the valley but from the Fedioukine Hills on one side and the captured redoubts on the other. By good fortune the French Chasseurs d'Afrique were directed on the Russian batteries on the former side, and took or routed them with great gallantry. Pennington and his fellow soldiers had heard the orders given by Lucan and Cardigan with incredulous amazement, 'for the madness of our errand was plain to the weakest judgment amongst us', but they rode on unflinchingly, every now and again obeying the shouts from officers and N.C.O.s of 'Close in, close in!' as more and more men and horses went down. Pennington lost his two comrades on his right and left, killed almost immediately.

Sergeant-Major Loy Smith of the 11th, riding just behind Private Young, felt that 'the very air hissed' with bullets and when a cannon ball took off Young's right arm was bespattered with the flesh. But Young only fell back coolly and asked what he was to do. 'Turn your horse about and get to the rear as fast as you can,' replied Loy Smith.

All was now confusion. When the 13th and 17th were some eighty yards from the end of the valley twelve Russian guns fired a salvo at them and the first line was more or less destroyed. Cardigan, almost removed by blast several lengths from the guns, was the first man into and through the battery, and then as the second line of the 11th on the extreme left outflanked the Russians the third, led by the 4th and finally the 8th, crashed in on the Russian guns. Fierce hand-to-hand fighting followed, and the guns were captured despite all Russian attempts to remove them.

The 11th, to the left of the Brigade, charged a body of Russian Lancers successfully, but were soon down to eighty men; now they found themselves facing a huge mass of enemy cavalry and infantry; this was at the base of the Fedioukine Hills. As Paget recorded 'it may be fairly said that on this occasion about forty men of the "Cherubims" advanced against the entire force of the Russian cavalry! Indeed, the Russian Army!' But in spite of the great odds, Douglas led his small force on, until carbine fire and the close attentions of Cossacks compelled him to retire, after he had got within forty yards of the main Russian body. Loy Smith believed that if some more British squadrons had then come up to help the Russians might have surrendered, 'for we, numbering now not more than eighty sabres, held this Russian Hussar Brigade in a corner for some minutes'. Looking round for support, Loy Smith saw

The Charge of the Light Brigade, Balaclava, October 25, 1854. The 11th Hussars are in the foreground, while the 17th Lancers can be

Army Museums Ogilby Trust

a body of Lancers forming up in his rear, seeming to have arrived from the Tractir Bridge direction. He knew they were Russians from their green and white pennants, but Colonel Douglas mistook them for the 17th and shouted 'Rally, men of the 17th Lancers!' Cornet Palmer told him 'It's the Russian lancers, sir,' and Douglas cried 'Then fight for your lives!' At this moment Paget, whose remnants of the 4th joined the 11th, rallied his own scattered dragoons shouting 'Halt, front! If you don't front, my boys, we're done!' By this time the scene was a hopeless mêlée with individual troops and even regiments hopelessly mixed up, but Paget and Douglas led some sort of an organised charge at the Russian lancers who had cut them off, and by sheer determination and courage got through with few casualties. In Douglas's words: 'If they had been pluckily handled, not a man of us would have escaped.'

Meanwhile it was left for every man separated or cut off to fend for himself and find his own way back. Many remarkable stories could be told by survivors. Riderless and wounded horses were a menace, and wounded or unhorsed soldiers were often cut down or shot by the Russians as they were trying to make their way back or catch another horse. Pennington tells of the luck of a colleague, Tom Spring:

> 'He fell with his horse after passing through the battery; and was unable to extricate his foot from one of his stirrup-irons, which was overpressed by his horse's dead body. He explained at this time that his sword was discoloured with blood, and that this sight may have kindled the cruel ire of his assailant. But a Russian officer [Tom thought of high rank] descrying him in the plight I have shewn, in the most dastardly manner fired every chamber of his revolver at the prostrate and helpless Hussar. It was only a month or two ago Tom shewed me the deep indentation from these bullets directed at his breast; any one of which

would doubtless have proved fatal, but for the resistance offered by the woollen padding of his hussar jacket.'

The toll of the canister or grape from the Russian guns had been terrible, not least on the horses. Loy Smith lost his charger as he was returning up the valley, and struggled on on foot with drawn sword, with at least half a mile between him and safety, expecting every moment to be hit by bullets from the infantry which spattered all round him. He saw two of his fellows cut down by Russian lancers eighty yards behind him, but miraculously ran on unharmed to the centre of the valley, where the bullets slackened off and he felt entitled to take a swig of rum from a small indiarubber bottle that he carried. Looking about him Loy Smith found that he was standing quite alone between the two armies about three-quarters of a mile from the battery that he had charged and a quarter of a mile from the British position. Encouraged, he walked on over a hillock where he found three Russians, one dead and two wounded; one of them who had lost a foot turned out to belong to the Russian 11th Hussars, a coincidence which so impressed Loy Smith that he nearly put paid to the man's heart by drawing his sword, only to cut off one of his buttons—which, ever after, he wore on his own uniform. Loy Smith managed to catch an unwounded horse of the 4th Light Dragoons, and with a Russian carbine in hand made his way back to 'J' Troop, R.H.A. and safety. But he was still burning with indignation at what had happened, and when some gunner friends hailed him only replied: 'Someone will have to answer for this days' work.'

Pennington too had had his troubles, and never reached the battery, as his black mare was struck in the hind leg by a musket ball comparatively early on.

'I can recall the sinking of my heart as I beheld the 11th ride on, while I was left alone far from the British lines to do the best I could. But I felt a strange reluctance to dismount. "Black Bess" had been the fastest mare in all the troop; high-bred and hardy, she had borne campaigning well. With no soul in sight, for the regiments in front had passed away obscured by dust and smoke, I made a good mark for musketry fire. A ball passed through my right leg, a shot from the left tilted my busby over my right ear, while "Bess" received the coup de grâce which brought us both to earth, though I was still astride the mare. My case seemed desperate, but there was no time for vain regrets. My leg was bleeding though I felt no pain. I looked around for refuge from their pattering fire, for bullets were still raising little clouds of dust quite near; and now I shuddered as I beheld disarmed and helpless men of ours singly attacked by many lancers who did not scruple to kill them in cold blood! I still could move upon my feet, and standing clear of poor dead "Bess" and unbuckling the waistbelt which encumbered me, held fast my sword, resolved if this best to sell my life as dearly as I might. I think I must have been distant a mile, or even more, from the point whence Lord Cardigan had given the orders to advance. My leg now giving signs of stiffness coming on, I had some half-formed plan of hobbling to the rear, when to my great surpirse I heard the thud of hoofs behind. They were the 8th Hussars! Great was my relief and joy as Sergeant-Major Harrison, seeing my plight, halted and bade me mount a grey mare he led as he rode serafile. Its rider had been killed. Lame as I had now become, I yet contrived to mount.'

T.S-M. Harrison, 8th Hussars, was nicknamed 'Old Bags', Pennington tells us, because he 'would wear his overalls loose and easy'. He was later commissioned.

Pennington then took part in charges that the 8th made against Russian squadrons, lost touch with them, but got back up the valley pursued by several Russian lancers, and finally reached the safety of the 'Heavies'. They had at first been ordered forward behind the Light Brigade, but when Lucan, who rode in between the two, saw that they were beginning to suffer casualties he sensibly halted them. Lucan himself, wounded in the leg, showed complete absence of fear, just as did Cardigan.

The latter, having ridden over a gun-carriage in one of the Russian batteries and received a lance thrust which did him no serious harm, galloped out of the smoke; 'Ronald', wild with excitement, carried him within twenty yards of a mass of cavalry. By extraordinary coincidence they were commanded by Prince Radzivill, who had met Cardigan in London. He now ordered some Cossacks to capture the Earl, but although cut by another lance thrust on the thigh Cardigan evaded them and galloped back.

The scene that now met his eyes and those of the watchers on the heights was a ghastly one. The Light Brigade, though it had disabled a battery of guns and slain and wounded a good many Russians, even taking a few prisoners, had ceased to exist as a fighting force. Scattered on the dusty plain lay bodies of men and horses, the wounded, the dead and the dying, and here and there survivors in tiny mixed groups or singly rode or walked their tired way back, still followed by sporadic Russian fire, or pursued by horsemen whom they had to fight off with carbine and sword, every now and again trying to catch and mount a riderless horse. To Cardigan, the destruction of his pride and joy was a deeply felt personal blow, the more so since he had felt the original order to be an impropriety.

Cardigan, who everyone agreed had led his brigade most gallantly, now did a curious thing; he made no attempt to rally the survivors, leaving that to Paget and his regimental commanders, but rode back, evading the Russians, along the way he had come, once he had got clear slowing his horse down to a walk. He might have been pondering the madness of the day and of his superior officers. As Pennington related: 'He certainly did not bring his Brigade out of action, but he led them nobly in. Lord Alfred Paget once said to me, "Cardigan took you in, Pennington, but my brother George brought you out." ' This was the truth. The circumstances of the charge and the return were however, to say the least, exceptional. At first sight the total casualties seemed terrible; out of 673 horsemen who had charged down the valley only 198 answered the roll-call. But many more turned up later, and though various figures are given by different authorities it seeems that the likeliest casualty total is between 247 and 300 all ranks killed, wounded and missing, and 475 horses killed or so badly hurt that they had to be destroyed. Some figures for human casualties are considerably higher, though that given is bad enough.

Back at the start line, according to Private Daniel Deeran, of the 4th, Cardigan rode up to the surviving Light Dragoons and said, 'This has been a great blunder, but don't blame me for it.' T.S-M. Frederick Short, also of the 4th, quotes him as saying, 'Men, it is a mad-brained trick, but it is no fault of mine,' and adds: 'I heard some of the men, who were naturally then rather excited, say 'Never mind, my Lord, we are ready to go again!' Lord Cardigan replied, 'No, no, men; you have done enough!', I heard no command given that day by Lord Cardigan while we were engaged; that is to say from the time we started until our return.'

Raglan, the real co-author of the crime, was furious with both Cardigan and Lucan, jerking the armless sleeve of his frock-coat, but Cardigan could rightly claim that he had only been following orders, and when Raglan cooled down he had to admit that Cardigan's conduct had been blameless. Lucan was not so lucky, for Raglan shifted or tried to shift the blame for the mistake on to him. It was one of those cases which could be argued for ever without a satisfactory upshot. Even so, when Lucan was sent home a fortnight later, nearly all the cavalry thought he had been hardly treated. As the historian Nolan remarked, Lucan had been 'as well he might be profoundly astonished' when he read the order Captain Nolan had delivered with such disrespect and inexcusable peremptoriness; W. H. Russell, who like Paget and Mayow remarked upon Lucan's indefatigable care for the cavalry, believed he had had as little to do with the ' "glorious" Balaclava bungle of the Light Brigade as I had.'

To the French General Bosquet it had been magnificent, but not war; the Russian General Liprandi thought the Light Brigade must be drunk, as well he might, and was surprised to find from prisoners that they were not. Liprandi asked 'Who was the General who went back on the chestnut horse with white heels?' and on being told that it was Cardigan remarked that he was lucky to get back, and never would have done if he had not had a good horse, so closely did the Cossacks chase him. Confusion between 'Ronald' and the very similar chestnut charger of Lieutenant Houghton, who, mortally wounded in the head by a shell fragment rode out of the battle still living, is the explanation for the affidavits of some soldiers that Cardigan had galloped away before the guns were reached—and for the subsequent canard that Cardigan had never taken part in the charge at all. The recriminations that followed in after years between Lucan, Cardigan, Airey, Paget and Lieut-Colonel the Honble. Somerset Calthorpe, 5th Dragoon Guards, were a blot on the battle and did none of their reputations any good, though Paget emerges as the most sensible, and Calthorpe would undoubtedly have lost Cardigan's libel suit had not the latter been non-suited on grounds of time elapsed.

Veterinary Surgeon Gloag, who watched the battle from a nearby hill, told his wife in a letter that the 11th lost seventy-two horses killed, thirty-two desperately wounded, of which many had to be shot. Because of sickness only seven officers took part in the charge; Colonel Douglas distinguished himself by leading and thrice rallying the regiment, had his mare struck on the neck but got away because of her speed, his revolver having been hit and exploded by a shell fragment; Lieutenant Palmer took a Russian Colonel prisoner at the guns; and Lieutenant Alexander Dunn, a Canadian, whom Pennington describes as one of the handsomest men of his day, and also as one of the finest swordsmen and horsemen in the Army, won the Victoria Cross; having emptied his revolver at the Russians he flung it at them and resorted to his sabre, which he used to such good effect (Dunn stood six feet three and used a sword much longer than regulations permitted) that he saved Sergeant Bentley's life by cutting down several Russians who were attacking him. Having had his own horse shot underneath him Dunn finished on one belonging to the 13th, and this one was also shot; Captain Cook was another of many whose horse was shot under them, and he and Lieutenant Trevelyan were wounded; the latter, having an exactly similar wound to Pennington's, 'behaved like the good samaritan and gentleman that he is, and urged me to partake of the contents of his haversack and flask.'

Balaclava is usually regarded as a disaster. In a sense it was: but as Paget rightly remarked, 'it really ought to be considered in the light of a victory', as far as the Light Brigade were concerned, since they attacked and disabled a line of eighteen field guns. The defeat surely lay in the utter failure of Raglan and the French first to prepare properly for the battle, then to exploit an overwhelming moral success—following up that of the Heavies—not forgetting the magnificent work of the 4th Chasseurs d'Afrique. Let Loy Smith sum up:

> 'It had been expected, and was well known that the Russians were preparing to attack us, for only two nights before we stood to our horses the whole night in the plain, about half way between Canrobert's Hill and our encampment, still no preparation was made. Had a few battalions of English or French Infantry been

posted in the Redoubts to support the Turks, and more of our Artillery brought into action, the day would have ended very differently. . . . We cut their army completely in two, taking their principal battery, driving their cavalry far to the rear. What more could 670 men do? A glorious affair might have been made of it, had our infantry been pushed along the Causeway Heights with the Heavy Cavalry, and the French Infantry with the Chasseurs d'Afrique along the Fedioukine Hills. The enemy were so panic-stricken, that I feel convinced the greater part of this army of 24,000 men would have been annihilated or taken prisoners, they having only two small bridges to retreat over, the Tractir and the Aqueduct. Never, I should say, was such an opportunity lost.'

Chapter 8

1854–1904

'... I understand it has been stated that the British cavalry are of a very inferior description, and require a thorough reform; that they are badly officered, being commanded by gentlemen of too high a rank in the country, and that they ought to be better officered. I can only say, that I do not think you will find any body of officers more careful of their men than those officers who now exist, and perform their duties better in the cavalry regiments, or that you will find any regiments in the world where there is such a mutual attachment between officers and men, as is the case in the British cavalry.' (The Earl of Cardigan, 1855.)

WHEN Cardigan had spoken to the survivors of the charge, which in all had lasted little more than half an hour from start to sad straggling return (between 11.35 and 11.45), and had ignored with a clear conscience Raglan's rebukes—later Raglan was to write of Cardigan's 'greatest steadiness and gallantry, as well as perseverance'—he returned to his yacht, where he treated himself to what some might think a well-deserved bottle of champagne. But not before siting his guns and posting a guard in case the Russians made a surprise attack, for detractors of Cardigan's yacht never mention that it was armed.

His regiment had lost at least twenty-six all ranks killed, twenty-seven wounded, and eight captured, (Gloag says thirty-two killed), and the wounds of some survivors were so frightful that death might well seem a blessing. Other regiments had suffered as badly, fifty or sixty per cent casualties being accepted. As Gloag two days later told his wife: 'Our numbers now on parade appear quite ridiculous. None of the Light Brigade have anything like half their numbers.'

At 11.45 the Russians had pulled back, but they still held the three eastern redoubts—from which the seven British naval guns had been removed in triumph—and more important, commanded the metalled Worontzoff Road and the Causeway Heights, recapture of which by the British infantry would have later spared the whole Army from much of the misery of the next winter, because of its vital importance for communications and supply, when the rain made the countryside a quagmire. The first general Allied advance towards the redoubts had not started before noon, after one o'clock cannon fire from both sides had dwindled, and by four o'clock ceased. The battle was over—a dubious draw from the Allies' viewpoint, although they had not lost Balaclava itself.

Of many curious stories relating to Balaclava one is especially pleasing; Lieutenant Dunn, V.C., talking to Lieutenant (later Lieut-General) Palmer the same night, said: 'You had a very near shave of it today, old fellow, as while we were rallying, after the first halt, a Russian came up behind you and put up his carbine close to your head. You did not see him, but Private Jowett charged him and cut him down.' Palmer, who when Orderly Officer a few nights before had found Jowett asleep at his post but had only cautioned instead of confining him, replied: 'Lucky for me I had saved him from getting flogged, otherwise he might not have been in such a hurry to save my life.'

Operationally the Cavalry Division was never to play any large part again in the war; how could it, indeed, when after Balaclava and with increased sickness among men and horses in the autumn, it was soon down to an operational strength of 200, instead of 2,000? In some ways, however, the worst part of the campaign was still to come.

The Allies, especially the British, were now feeling the pinch of numbers, whereas the Russians had reinforced Sebastopol to a total of 115,000 men against their 65,000 opponents. The eastern flank of Raglan's Army was dangerously exposed; nevertheless, with the French under Canrobert at last ready to attack, it was decided to assault Sebastopol on November 7. It was too late: the Russians, who had already made one unsuccessful sortie on October 26, launched a full-scale attack on the Inkerman ridge overlooking the Tchernaya in fog early on November 5. The bloody and swaying battle that followed was almost entirely an infantry affair, in which many French and British regiments distinguished themselves. Inconclusive in result, it was yet a numerical and brave victory for the Allies. The Light Brigade survivors under Paget were shelled heavily but made only one diversionary advance with the French Chasseurs.

From this time on the lot of the troops became worse and worse as storms, rain, mud, lack of food, supplies and even clothing, and finally cold beset them. Administratively the Balaclava base and supply arrangements were disastrously inadequate and incompetent. The cavalry had been moved seven miles from the base to a plateau near Sebastopol, utterly exposed, with no blade of grass for the poor horses. With the breakdown of fodder arrangements and the clogging with mud of the one available road to Balaclava, they became too weak to walk to the base and return laden, so that a great many died—thirty-five in one twenty-four hours' spell. Despite Raglan's wish to move them back nearer base Canrobert asked particularly that the cavalry stay where it was because that would prevent the Russians attacking again.

All this time because of casualties reinforcements had been dribbling in from time to time; more men were being lost from sickness than from enemy action. Private William

Ellis wrote, 'We remained at Inkerman till the horses and men were nearly starved to death. The horses hadn't a bit of mane or tail left, and there wasn't a strap left on the saddles—the horses having eaten them!' And so it went on; on Christmas Eve fifteen horses in the regiment died. In six months of winter, 1854-55, 439 out of 1,161 horses in the light cavalry died from starvation or sickness.

All this could not be laid at the door of either Lucan or Cardigan. At last Balaclava, if only briefly, had found them on common ground, and for a time after the battle each exculpated the other from blame. Raglan, however, was still upset by Lucan, and though, Airey told Paget, he admitted that, deplorable as the affair had been, it would have some results and make a great impression on the Russians, he soon engineered Lucan's dismissal. Cardigan, a sick man, had already left for good early in December 1854. But at least both men were expert with horses; it was the staff and the commisariat which were to blame for the cavalry's miseries, not their own generals. Mrs. Duberley, the spirited wife of the paymaster of the 8th Hussars, who had by no means always been an admirer of Cardigan, recorded when he was invalided home: 'I was discoursing with him yesterday, and he said: "My health is broken down—I have no brigade—if I had a brigade I am not allowed to command it. My heart and health are broken. I must go home." Ever since he has been in the Crimea he has behaved very well, and upon my word I'm sincerely sorry for him.'

By the end of November many of the camps were 'literally morasses', in E. H. Nolan's words; soldiers lived, ate, drank and slept in the mud. Now cholera struck again, carrying off sixty a day in the Army, and other diseases and fevers too. The medical situation was appalling. Russell recorded that on January 2, 1855, there were 3,500 men sick in the British lines before Sebastopol from bad weather and exposure. Conditions were pitiable, with 'no boots, no great coats—officers in tatters and rabbit-skins, men in bread-bags and rags; no medicine, no shelter. . . .'

As Russell put it: 'We have out here "soldiering with the gilding off," and many a young gentleman would be for ever cured of his love for arms if he could but see one day's fighting and have one day's parade of the men who do it.'

Not the least evil was the total absence of fresh meat, which meant that for months on end the soldiers had only salt meat. Even the cholera, which the Tulloch-McNeill Commission later stated caused only 1,200 deaths that winter, could have been endured as 'act of God'; whereas the 8,800 deaths from diseases which could with proper attention have been avoided were inexcusable.

At last the horrors of the winter passed, and from the spring of 1855 conditions improved all round; even material for making huts arrived. On March 2 Czar Nicholas died; in June the Allies took two Russian forts but a general assault on Sebastopol, at which the 11th were present, failed with heavy casualties on June 18, and ten days later modest, intelligent, long-suffering, brave, and in many ways disastrous Lord Raglan died. In early August the Russians made a desperate attack on the Allied rear but were beaten off with severe losses (8,141) by the Sardinians particularly and by the French, supported by the remnants of the Cavalry Division, now under Scarlett.

The final attack on Sebastopol was made on September 8, after three days' terrific bombardment. The French, since May under the energetic General Pélissier *vice* Canrobert, brilliantly stormed the Malakoff, the British bloodily failed to capture the Redan; all the same the Russians that night completely evacuated the southern side of the city and harbour after blowing up their magazines and bridge. From then on the war drifted futilely to its end, and on March 30 1856 peace was signed. The second winter had been spent by the 11th in comparative comfort at Kadikoi. But whereas in 1854 they had lost four officers and fifty-eight men dead from action and sickness, in 1855, with far less action, their casualties had been one officer and fifty men dead. In all, in two years, the regiment had lost five officers and 109 N.C.O.s and men 'Died in the East'—how reminiscent is the *In Russland Gefallen* of Hitler's War—eight prisoners of war, and thirteen officers and seventy-two N.C.O.s and men invalided home, some of them, alas, to die also. From Raglan's Army 1,407 officers and 14,901 men had been invalided home. And the dead, in battle and through disease or terrible conditions? 264 officers and 22,187 other ranks. Of those, nearly five times as many died from the latter as the former cause. The sickness figures tell the story; April 1854, in Turkey, 503; July 1854, Varna area, 6,937; September 1854, 11,693; December 1854, 19,479; January 1855, 23,076!

The war had resolved nothing usefully, and scarcely a leading general or an admiral had enhanced his reputation, unless Vice-Admiral Korniloff. Many, like him, had paid with their lives, others with professional or public esteem in a graveyard of reputations. Perhaps only the German engineer Todleben had survived gloriously improved in deserved repute.

Yet there was one other whole class which, having suffered brutally, had emerged to new prospects almost literally through mud and blood. This was the British private soldier, who had in the cavalry as in the infantry and artillery displayed unconquerable resolution, nerve, and discipline. In the words of Mrs. Woodham-Smith, 'At the beginning of the campaign the private soldier was regarded as a dangerous brute; at the end he was a hero.'

At last long overdue reforms in welfare, medical and supply services, punishments, education, and not least in the requirements for officers, were in sight, even if, just as in Marlborough's day, the crippled or worn out veteran still awaited generous pensions in vain and pay itself continued derisory for most. It was not for nothing that the exhausted cavalrymen of the Light Brigade had given three cheers when they had ridden back from the Valley of Death.

The regiment arrived back in England in late July 1856, and were promptly inspected by Queen Victoria, and later by Cardigan himself, now promoted Inspector-General of Cavalry, in which post he became the terror of every regiment that he visited. For the moment, however, he was the most popular soldier in England. In 1860 he achieved his life's ambition—apart from the elusive Garter—by becoming Colonel of the 11th, or 'Cardigan's Bloodhounds', now thought a most suitable term for them.

Cardigan had been awaiting this honour since the review in Hyde Park on June 26, 1857, before Queen Victoria and the Prince Consort, when the Queen, dressed in a scarlet riding-coat and wearing a general's sash and plume, presented the first Victoria Crosses. The 11th were amongst those on parade, and the regimental slow march, *'Coburg'*—traditionally composed by an aunt of Victoria's—was played during the march past in slow time.

In 1858, after the death of his first wife, from whom he had long been separated, Cardigan married again, his second Countess being spirited, loyal, and somewhat eccentric, certainly in the eyes of the Queen. Mrs. Woodham-Smith tells us that the second Countess was addicted to Spanish dancing, daring clothes (she sometimes received visitors at Deene Park wearing crimson trousers after her husband's death) and hunting and racing society. Nevertheless Cardigan was promoted Lieut-General in 1861, by which time he was devoting, as usual when possible, most of his time to hunting in Leicestershire and his own Northamptonshire; in this country—and then, perhaps, even more than now, High Leicestershire of all others demanded great determination of both horse and rider, as Whyte Melville reminds us—his exploits in the hunting field became legendary. He took many falls, some crashing, in his stride; it was ironic that on March 27, 1868, when riding a young horse, out to see the bereaved family of a former tenant near Deene, Cardigan was thrown and crushed by the restless animal which had reared and fallen upon him. Whether he had a seizure or not, Cardigan never spoke again, and died next day.

During the last decade of his life Cardigan's reputation had soared to unmerited heights and then plummeted to undeserved depths. That his own apparent boastfulness and unquestionably fatal attraction to litigation were largely responsible for public disesteem, cannot be denied. That was only part of the story, however. He was a contentious man, who, moreover, simply did not care much about his reputation; as Whyte Melville says, 'few have sustained, it must be admitted with considerable indifference, so unreasonable a load of inconsiderate praise and unmerited censure. . . .'

Pennington, who while a romantic was no fool, thought Cardigan unfit for independent command in the Crimea because of his 'lack of judgment', but at the same time execrated Kinglake, the historian, also far from foolish, for his 'unwarranted innuendoes' against Cardigan. Pennington's advantage as a witness of human nature is that he served as a private under Cardigan and knew what the men of the regiment thought. He certainly knew what they would have said to Kinglake had the latter been derogatory about Cardigan in their presence. Pennington's verdict on his former Colonel is this:

> 'He possessed little mental capacity, and owed his military position to the power of wealth. He was one of the results of the unjust and degrading system of purchase. . . .
>
> 'He carried pride of birth and position to the point of snobbery, and his nature was by no means lovable. He was disliked, and, as a consequence, has been greatly disparaged by officers, but I never heard any of the

"rank and file" speak of him, as a soldier, in other than admiring terms. . . .'

E. H. Nolan notes that Cardigan had the reputation amongst the men of being 'a soldier's friend', and if Melville and others are to be believed this was equally true at Deene Park about his tenantry. George Ryan wrote in 1855 that public opinion, which knew all about Cardigan's reputation as a martinet,

> 'knew nothing of his many acts of kindness it was his study to keep secret. . . . Not only was he ever ready to minister to the pressing necessities of the married men, but many officers were in the moment of difficulty succoured by his lordship. The Earl of Cardigan's career has been daubed dark as night. His good deeds have had no public sitting, or their fair light would have materially toned down the harshness of the picture.'

Time also softens most things. In any case, one need see no reason to doubt his much younger widow's statement that she was 'ideally happy' with Cardigan, who 'was full of the joy of life and entered into everything with the zest of a young man'. For better or worse he had been immediately identified with the regiment for thirty years, and, as some may think, if the 11th could withstand, endure and even thrive upon Cardigan they could withstand anything.

At all events, when he was buried in Deene Church on April 9, 1868, although the regiment was now serving in India there was a tremendous turnout of 11th Hussars, and eight of his old officers carried the coffin.

It was the end of another epoch. But Cardigan, as always, seemed to have won in the end. For even today, one hundred years on, the regiment is the only one in the British Army that sounds the Last Post at ten minutes to 10 p.m. instead of on the hour exactly. The reason why? That was the time at which Cardigan died.

★ ★ ★ ★ ★ ★

The Crimea had taught many lessons, not all of them military ones, and some, at least, had been appreciated in Whitehall. War is a great leveller, but the bridgeless social gap between officers and men in peacetime was to continue into the next century. Apart from their soldiering duties, the two lived, throughout the Victorian and Edwardian ages and indeed until after the Great War, in almost totally different worlds. Nevertheless, changes for the better in social conditions in the Army and correction of some of its more monstrous inequities now began to be made successfully. As always they took a long time with a great deal of resistance to be overcome.

It had not been until 1852, for example, that married quarters had been sanctioned officially, and it was years before they became general in a form which could be recognised as such today, even as a caricature. The first *barracks* had only been built sixty years before, with the 'married corner' in some barrack-rooms, and over another half century had passed since 1792 before a barrack-room or rooms were allotted by units to married families. In the case of the 11th this did not happen until 1848, and they seem to have been one of the earliest regiments to have adopted the practice. Milton undoubtedly was correct in telling the Lord General Cromwell that

               'Peace hath her victories
        No less renown'd than war',

but some of them in England have taken a long time gaining.

The system of purchase and patronage of officers' commissions, exchange and half-pay, so longstanding and in many ways obviously abused, has been mentioned before. The Crimea particularly showed up how unfit many officers at every level were to command, apart from the fact that of the Generals most had never heard a shot fired in action. That could hardly have happened in a continental army, or indeed in the Indian Army, which provided some outstandingly successful officers at junior level. With all its vices, however, it is arguable that the traditional British system had achieved one good result; although it had been commonplace for some of the great English captains such as Marlborough to change sides politically, or to modify their political views, as in Monk's case before him, since the execution of Charles I there had always been great repugnance, in the country and Army alike, to any direct intervention by the Army *as a class* in constitutional crises. The Civil War had left bitter wounds and scars, Cromwell's Major-Generals by their arbitrary and sometimes police-state methods had alienated Parliament, whereas Monk, seeing that thereafter the Restoration of Charles II was inevitable, performed his greatest service by bringing it about without another civil war. As Sir Charles Firth has written:

> 'His dexterous and unscrupulous policy blinded the Republicans to his intentions till it was too late for them to resist, and made the army instrumental in effecting what the bulk of it would have fought to prevent.'

In the eighteenth century, despite some curious military-political goings-on, there was never any real danger of the British Army's becoming, as the Prussian Army did, a State within a State, nor in the nineteenth. For generations now this political uninvolvement has been a blessing here; a curse in some other European countries, of which Germany remains the most obvious example, has been its reverse.

By about 1860, Gronow relates ('but now, thank Heaven, our system is much improved') patronage could no longer do everything in the Army, and 'a strict examination is necessary for all candidates for commissions in the army'. Yet a few years before, at the end of the Crimean War,

Nolan contrasting the British with the French system, could write:

> 'Without fortune and aristocratic influence, a British officer, however well educated and brave, has no hope of rising to eminence. . . .
> 'The British purchase and half-pay arrangements, the beginning and end of our system, form a curious exemplification of the commercial spirit in military affairs—a strange combination of the Court and the Stock Exchange.'

Prospects for promotion were still not too bright for impoverished officers—but they were certainly not as bad as they had been during the previous century. Thus when Lieut-General Campbell had reviewed Ancram's Dragoons in 1755 the two senior subalterns, Lieutenants Alexander Stewart (Stuart?) and George Whitmore, aged fifty-three and fifty-five respectively, had served between them seventy-eight years in the army, almost all of them in the regiment.

Throughout the nineteenth century few ordinary soldiers could save much money from their pay, and although their living conditions improved gradually far too much of the private's money was 'blown' in the canteen. And the lot of the veteran of the Crimea or Indian campaigning, which might have crippled his health, was all too often as dismal as in Marlborough's day, when he had to leave the army and a grateful government accorded him a pittance. The story of Sergeant Richard Brown, who had served with the Light Brigade all through the Crimea, is a typical case in point. Brown was a friend of Pennington, to whom he gave in 1886 Colonel Douglas's sword, worn at Balaclava, writing to him, 'I know you will value it, and keep it in memory of him and your old comrade,
Richard Brown, (Late Sergeant Eleventh P.A.O. Hussars).'

Brown died towards the end of the century, and Mrs.

Kelly, who had talked to Pennington about him, wrote in 1902:

> 'He had been the favourite orderly of Lord Cardigan, and was for some years the devoted and trusty henchman of Colonel John Douglas. Handsome and honest, he was truly a model soldier, for, in his long service of twenty-one years, he was never in the defaulter's book. It was known that if he had not been illiterate he would have borne Her Majesty's commission. He certainly had a pension of the heroic sum of one shilling and three pence per diem, and for twelve years subsequent to his retirement from the service he worked (often ankle-deep in water) at a canal side in Manchester, but when age and rheumatism rendered him incapable, he was compelled to go to the workhouse. . . .
>
> 'Sergeant Richard Brown, forgotten by his country, died in the workhouse, and yet it has often been said of him that "no better man ever drew the breath of life". How often he must have regretted that he had not died with his comrades in the fatal North Valley, instead of having to look forward to filling the grave of a pauperhouse in his native land.'

Brown had been particularly unlucky, because a friend had tried to get him a job as a War Office messenger, but the Sergeant had been considered too old according to precedent and unable to adapt himself to the routine.

  ★ ★ ★ ★ ★ ★

Between 1857–65 the regiment was in England or in Ireland, where perhaps the most notable event that took place, affecting each of the eight troops and roughly 640 men in the regiment was the death in 1862 of 'Old Bob', aged 33, the oldest troop-horse in the British cavalry. 'Old' or 'Crimean' Bob had joined the 15th, then the 14th, then the 11th Hussars in 1838, and, assuming him to be a four-year-old when he joined, he was thirty-three when he died. He spent almost a quarter of a century with the regiment, went all through the Crimea, including the Charge, and was never once struck off duty because of sickness. The Duke of Cambridge, C-in-C, refused to allow Old Bob to be cast in England or Ireland, and when he died at Cahir the Farrier-Major who had ridden him all through the Crimea supervised his burial with full military honours.

From 1866–77 a long period of Indian service followed, where the regiment was lucky to have two fine Commanding Officers, Lieut-Colonels Charles Fraser, V.C. (from the 7th Hussars) and Arthur Lyttelton Annesley, both later Lieut-Generals. Cardigan died in 1868, so too by a gunshot accident in Abyssinia did Brevet-Colonel Alexander Dunn, V.C., who had used General Wolfe's sword at Balaclava to such good effect. He was only thirty-four.

In 1871 Dunn's old commanding officer, Major-General John Douglas, followed him to the grave. He had been nearly forty-two years in the army, and was Inspector-General of Cavalry.

In India the regiment was not just showily mounted—at reviews its two bay squadrons divided by a squadron entirely mounted on greys drew attention—but was very well trained by its Colonels, and drew glowing inspection reports, praise for its scouting system on exercises, and, perhaps most significant of all, tributes to 'the cordial good feeling which prevails throughout all ranks, and the almost total absence of crime' coupled with 'the high state of efficiency and discipline'. Credit for some of this must go to Lieutenant and Adjutant St John Taylor, who only gave up the position he had held for eleven years on his promotion to Captain in 1876! Relations with Indian cavalry regiments were so good that the Indian newspapers mentioned them. And when in 1879, a year after their return to England, the Inspector-General of Cavalry inspected the 11th, he told them: 'I have seen several regiments return from India, but yours is the best I have seen.' Peace indeed has its victories.

A new role: The 11th depicted as the villains in an American play of the late 1890's.

So does war. The death in 1881, aged ninety-six, of Captain George Ridout, who had joined the regiment at fifteen in 1801, almost recalled a different century. When he went to the Peninsula the regiment was 800 strong and each troop had horses of a different colour; he commanded a squadron at El Bodon, charged ten times, had two horses shot under him, and was hit by a bullet which was deflected by the Bible he carried in his pocket.

In 1884 the 11th lost another distinguished officer whom the Army could not afford to lose. This was Lieut-Colonel J. D. H. Stewart, C.M.G., C.B., General Gordon's Military Secretary, formerly second-in-command of the regiment. Stewart, a close friend of both Wolseley and Kitchener, had been ordered to leave Khartoum during the siege by Gordon on a special mission. He left in a steamer down the

Nile with the French vice-consul, a correspondent of *The Times*, and an interpreter, but was betrayed and murdered, with his European companions, by Arabs who had induced the party to land. In a letter written from Gakdul on January 12, 1885, Kitchener wrote:

> 'I think it was the 23rd September that I first heard that Gordon was going to send Stewart down. I immediately sent off special messengers to catch him at Berber advising him to take the desert track from there and warning him about the Robatab and the Monassir tribes....
>
> 'I served with Stewart in Anatolia and saw more of him in Egypt, he was a dear friend of mine and the finest soldier I have ever met.'

Elsewhere Kitchener described Stewart as 'the best officer in the British Army', and Wolseley was of much the same opinion. Stewart, who fought with his fists against swords, was only forty when murdered.

For almost seven years between 1892–99 the regiment was again in India. This time its record was not good, although it served its time during the 1897–98 Frontier troubles. Something appeared to have gone wrong with morale. The 11th had gone to South Africa between 1890–92, and on the occasion of their going the Prince of Wales (later Edward VII) had told them of the regiment that 'he was born in it, when my father was Colonel of it'.

Whether the regiment assimilated some curious ideas from the Prince about discipline, or not, a future and great 11th Hussar, Major-General T. T. Pitman, was to write about his experience as a young officer in 1889:

> 'I believe it was looked on, at that time, as the smartest regiment in the Army.'

But, General Pitman (writing in 1933) goes on, the average officer knew 'extremely little' about the art of war. Colonel (later Field-Marshal) French of the 19th

Hussars was thought at Aldershot to have a 'mad craze' for soldiering because he paraded his regiment mounted daily! As for the relationship between officers and men, 'the comradeship of to-day was unknown', and the whole of weekly pay found its way into the canteen.

Major Edwin Milson, a private when Pitman joined but later one of the great 11th Quartermasters, told him of that time:

> 'If you had come down of a night and suddenly ordered the squadron to turn out, there would not have been half a dozen capable of appearing on parade, whereas today there would be few absentees.'

Most officers, General Pitman recalled, did not even know the names of all the men in their troop—himself included; but he managed to bluff an inspecting general by inventing names for those whose real ones escaped him, and discipline was so good that no one betrayed him.

In 1892 in Sialkote, India, the 11th received 'one of the worst inspection reports that the regiment has ever had'. All leave was stopped for a month, and though the commanding officer, at that time Lieut-Colonel Charles Swaine, naturally bore the brunt, everyone felt the blame keenly. Action was taken.

For over two years, between 1890–92, the regiment had been in South Africa. Whether there was some grievance against it or not, by 1894 things seem to have sorted themselves out in India. The comments of the Duke of Cambridge, still C-in-C British Army, on the inspection report at the end of 1894 are these:

> 'H.R.H. considers the report very satisfactory, the improvement that has taken place in the regiment since the last Inspection being most creditable to Lieut-Col. Swaine and all ranks under his command.'

In 1899 the 11th, still in India, were first on the roster for war service. War did come, in South Africa, on October

12. But an outbreak of glanders among the horses on June 10 had put the regiment in quarantine for four months, and it had lost its place. All the same, 108 N.C.O.s and men, under Captain Esmé Harrison and Lieutenant P. D. FitzGerald, left in September as a dismounted contribution. They were all at the siege of Ladysmith, as were three more officers of the 11th serving with other regiments. Major Harrison's letters thence tell of mules and horses being eaten.

> 'They say there are hundreds of wagon loads of jam and nice things waiting for us with Buller. Roll on Buller!' (February 22, 1900.)

Harrison and FitzGerald were two of the eight officers of the 11th who won the D.S.O. during the Boer War.

At the end of October 1899 the regiment reached Suez, and for the rest of the Boer War Egypt was to be its base. This was disappointing, but Lord Cromer insisted on keeping one well-trained cavalry regiment in Egypt. Much of the credit—with the lessons of India still in mind—must go to one of the best commanding officers the 11th ever had Lieut-Colonel Honble O. V. G. A. Lumley. Lumley's position must have been frustrating, for he was never able to command his regiment in action. But the training, even inspiration, which he brought to his job has been reflected later in generations of the family of the Earls of Scarbrough.

The record speaks for itself. In 1901 a draft of eighty was sent to South Africa, fifty of whom went to the 8th Hussars; in 1902 another draft of 138—all to the 8th Hussars. Roughly half the regiment, including reservists, fought in the Boer War, or as General Pitman puts it, nearly all the officers and many of the men. The lessons learned were to be invaluable in the Great War.

In 1868 Enfield carbines had been replaced by Snider Carbines, breechloading. In 1902 the Lee-Enfield magazine rifle was issued. It is not hard to see why. Because the 11th had always had to take, and had taken, a keen interest in shooting, their performance was noted by the Army Rifle Association in 1903. Since 1896 the regiment had won the Queen Victoria's Cup for Cavalry five times—

> 'perhaps the most striking success . . . this year they have carried off all three events for which it is possible for cavalry to enter. . . .'

An in 1904 for the second year the 11th were the best signalling unit in the cavalry.

Chapter 9

The Great War: 1

'It was a fine force, the Cavalry Corps, for it still contained a good proportion of well-trained officers and men and possessed a full measure of great fighting traditions. To the officers and men of such a force it was almost intolerable to see splendid infantry battalions marching to the front and to hear of them later coming out of battle almost annihilated, while the cavalry remained standing by their horses in billet or bivouac.'

(Captain L. R. Lumley, *History of the Eleventh Hussars (Prince Albert's Own) 1908-1934*, writing of the period after 2nd Ypres (May 1915) until November 1917.)

IN 1897, when there had been a sudden flare-up of trouble on the North-West Frontier of India, there had been some hurried and yet evidently slow-moving mobilisation orders issued by the British Army. The 11th, stationed at Sialkote under Lieut-Colonel E. R. Courtenay, realised that something might be in the wind when he, having just gone on leave, sent a telegram announcing his immediate return. However, the first unofficial news of their destination came in another telegram next morning (how typically, even fifty years later!)—from the coffee shop wallah. It read:

'From Ebrahim Bux. Hear Regiment ordered Peshawar. Can I have management of Coffee Shop?'

The lessons of India and the Boer War in this sort of regard—mobilisation especially—were not lost on officers such as Pitman and Milson, nor, fortunately, upon some in higher places in the War Office. By 1912 the 11th, with some forethought, had perfected a mobilisation scheme worked out by these two (Milson had done most of the spadework and published a pamphlet on the subject in 1911, and Pitman took over command of the regiment in September 1912). This scheme went up to the Horse Guards eventually, and was, apparently, copied very closely by all other cavalry units. Practice exercises perfected theory, so that when the 11th went to war in 1914—and a large number of reservists were called up—everything went like clockwork. This would have impressed their Colonel-in-Chief, had he been able to witness mobilisation. But he was 'Little Willy', H.I.H. William, Crown

Army Museums Ogilby Trust

Crown Prince Wilhelm of Germany, son of the Kaiser, with the 11th Hussars at Shorncliffe in 1911.

Prince of the German Empire and of Prussia, whom King George V had appointed the regiment's first Colonel-in-Chief in 1911, shortly before his own Coronation.

The regimental journal, which had published its first number in April 1910, had this to say about H.I.H.:

> 'His *bonhomie*, geniality, and willingness to disregard the *convenances* which hedge in Royal personages, have made him a universal favourite with everyone he has come in contact with.'

In June 1911 'Little Willy' reviewed the regiment at Shorncliffe, and afterwards played for the regimental polo team against the Shorncliffe garrison. He was on the winning side (8–4).

The Old Comrades' Association had been founded in 1909, and the first Old Comrades' Dinner held the same year. When Major Pitman had approached some older officers who had left the regiment decades earlier a typical response to his suggestion had been:

> 'Do you mean that officers are going to sit down at the same table with the men? What sort of respect will you get in barracks the next day?' He replied: 'Double the quantity.'

Later, General Pitman was to say that 'A hundred times' would have been a more accurate estimate.

These innovations were typical of sensible and effective measures taken in the 11th to improve relations between officers and men. The regimental journal might seem a small contribution, but in reality it soon became—and has always remained—a most valuable one. By the time war broke out in August 1914 not only had the immensely improved training of the previous twenty years had great effect, but there existed also a noticeable movement towards real comradeship between officers and men. A beginning only, perhaps, but war, as always was to give it huge impetus, so that soon it became unstoppable.

It is well to emphasize that in size a cavalry regiment has

usually been only the equivalent of one large infantry battalion, often less than one. Despite cross-postings, drafts, transfers, retirements, resignations, reservists, and everything else common to most other arms of the service, it is essentially a small unit. It has the great merits and sometimes the weaknesses, certainly the disadvantages, of small units. But its spirit is perhaps harder than that of most to destroy, perhaps because the numbers of those who have served happily in it at one time or another far exceed its nominal strength.

In 1908 and 1909 the 11th won the Inter-regimental polo tournament (they had been trying to do so for thirty years), and in the jumping ring they did well also in those glorious Indian summer years of Imperial Britain before 1914. Captain (later Major) R. M. Stewart-Richardson trained the jumping team, and in 1912 and 1913 himself represented Britain in a team of officers at the New York Horse Show riding for a large part of the latter meeting with a broken ankle without stirrups. Stewart-Richardson was to win three Military Crosses in the Great War, and it was his previous horsemastership that helped many of the regiment's horses during it.

In musketry the almost unique record of many years past was being kept up, and if fewer prizes were won, this was because the standard of competitors had risen and not because the regiment's had fallen. In 1914 the 11th went to war without a single third-class shot, nearly four out of five men being either Marksmen or 1st-class shots. Much of the credit for this was due to Lieutenant E. L. Spears (later Major-General Sir Louis Spears, C.B., C.B.E., M.C.) and S.S-M.I.M. T. G. Upton. The latter had joined as a private in 1904, and served continuously with the regiment until 1932 when he became Quarter-Master at Sandhurst. When he retired, as Lieut-Colonel Tom Upton, O.B.E. D.C.M., he had been through just about every rank open to him in the regiment.

General Spears, who had joined the 11th from the 8th Hussars in 1910, has written in *The Picnic Basket* about officers and men in those far-off days when the sun always seemed to shine brighter and longer in the summers, the winters to be less cold and depressing.

'Everyone felt intense pride at being a member of this splendid Regiment mounted on such fine well groomed shiny-coated horses, from the Colonel at the head with his trumpeter behind him, the Adjutant at his side, down to the young trooper on his first route march. To the onlooker it seemed quite an army, the three squadrons each led by its squadron leader, also followed by his trumpeter carrying the silver-noted trumpet for dismounted calls and the shrill bugle to convey orders through the din of galloping hoofs; and the regimental headquarters, the equivalent of another squadron, which included the two machine-guns of the Regiment and their teams. Twelve troops there were, each led by a subaltern or his troop sergeant. . . .

'The non-commissioned officers should have their share of any description of the Regiment, for if, by extreme licence, we, the officers, could be described as its intellect, they were certainly the bones. Powerful men they were, and vigorous, sometimes martinets, but often harbouring behind the mask of strict discipline a sense of humour best expressed with a straight face. . . .

'But of this splendid and completely reliable band of men there is one I shall not only remember with pleasure but shall always consider I was both honoured and fortunate to work with: Sergeant Tommy Upton, my musketry instructor sergeant. Elegant, fine boned and clear eyed, he was a typical and very smart light cavalryman. He was also ideally suited to his job, keen and painstaking. We used to sit for hours in my room trying to devise means of making musketry interesting to the rank and file and to circumvent the contempt in which it was held by some officers. We were not interested in trophies or cups, leaving the marksmen and the pot hunters to look after themselves. He finally became a Major. There was no mission requiring courage, no post demanding initiative, no task requiring hard work he would not have distinguished himself in. So long as we, as a nation, can produce men like Tommy Upton we have nothing to worry about.'

It was not Spears's lot to serve in the regiment in the Great War, but his amazingly successful career as liaison officer with the French (he was wounded four times, so it was hardly a sinecure from the purely military side) and his subsequent distinctions never blinded him to the vicissitudes and achievements of his fellow 11th Hussars; even today he is one of the most loyal members of the regimental family. Colonel Upton would have been delighted to read in a 1969 list of promotions his son Peter's gazetting to Lieut-Colonel after nearly twenty-five years' service as an 11th Hussar officer.

While the intensive and varied training went on between 1908–14 the regiment had to find frequent drafts for the 13th Hussars in India. In 1910 no less than 166 N.C.O.s and men were sent out, some of whom helped the 13th add to its fine reputation in later cavalry operations in Mesopotamia. Training is a continuous as well as a progressive business, and one should go further back still when contemplating some of its results during the Great War. For between 1914–18 eleven 11th Hussar officers, past or serving, became Generals in various capacities, and several commanded other cavalry regiments or infantry battalions. Over thirty N.C.O.s and men received commissions during the war, three of them in the regiment; and nearly twenty who had served with it before the war obtained commissions in other regiments or corps.

Of the former, five had already won the D.C.M., and two added an M.C. to it (Captain J. P. Howells with The Royal West Surreys, and Lieutenant A. G. Clifford with The Lincoln Regiment); L./Cpl H. Gough went on to win a D.S.O. with The North Staffordshire Regiment, Cpl E. J. Richardson an M.C. with The Norfolk Regiment, and R.Q.-M.S. W. M. Lummis an M.C. with the 2nd Battalion The Suffolk Regiment; two others became Majors. Captain Lummis for many years has been better known as Canon Lummis, a great archivist and historian of both his regiments.

Of the latter group former Sergeant E. R. Orme also won the M.C. and commanded the 22nd Battalion Royal Fusiliers; W. T. Calvert won an M.C. with the R.F.A.; and former S.S.M. E. T. Sayer, who won the M.C. with the Essex Yeomanry after being severely wounded in 1916, joined the R.F.C. in 1918, and after a varied and distinguished career in the R.A.F.—in which he became known universally later on as 'Grandpa'—wound up as a Wing-Commander and finally as Secretary, R.A.F. College Cranwell in 1930.

Writing of the original B.E.F. of 1914, General Pitman thought: 'The cavalry soldier was a long way ahead of his infantry confrère. He had been trained to think in miles and not in feet.' There was much greater decentralisation, for all from Colonel to Corporal had their own unit to lead, and the troop leader, usually an officer, but sometimes a sergeant, was entirely responsible for his troop, 'and in the field would be sent off over the skyline with nothing to guide him but a map and his own wits'. Whereas the infantry subaltern 'was still chained to the apron-string of his company commander, and was seldom out of his sight'.

All this, from the infantry's point of view, was to change drastically and often in tragic circumstances—and rapidly too—especially after 2nd Ypres when the terrible semi-static war of attrition started that was to last so long and so bitterly in often appalling conditions. The irony and the frustration were from the cavalry's point of view that as any further war of movement became impossible they could never again fulfil their proper function as horsed cavalry. This is not the place to argue the rights and wrongs of keeping them mounted. However obvious it has seemed for many years that French and Haig made the wrong decision, the cavalry regiments themselves cannot be blamed for what happened or for not having to endure the constant hell of the infantry. Their own doses of hell dismounted in the line were far less frequent, but they swallowed them just as bravely and performed whatever job they were given quietly, efficiently, and sometimes gloriously.

One thinks of Spears making every man in the regiment pass through a special musketry test which he conducted personally, with Upton at hand; of Stewart-Richardson training the horses; of Captain A. B. Lawson, Adjutant until just before the war, and in 1914–15 the regiment's finest squadron leader, poring over mobilisation plans and strengths.

In June 1913 Lieut-General Sir Douglas Haig had inspected 10,000 troops, including the regiment, on Laffan's Plain at the King's Birthday Review at Aldershot. Symbolically, it now seems, the R.F.C. made its first appearance in public and twelve planes flew past the saluting base. It was also symbolic that the King and Queen watched the 11th make a dismounted attack in 1913. Haig, C.-in-C. Aldershot, a soldier never given to lavish praise, wrote on the regiment's last annual inspection report before the war:

> 'This is a fine regiment, well commanded and well trained. The high standard reached in musketry is particularly satisfactory. There is also a cheerful soldierly spirit in the Regiment.'

On August 15, 1914, the regiment embarked at Southampton for France and the war whose course no one could rightly predict, confident as many people were that they could. It was commanded by Lieut-Colonel T. T. Pitman, with Major R. J. P. Anderson, D.S.O., second-in-command; the three sabre squadrons were led by Major W. J. Lockett, D.S.O., ('C'), Captain A. B. Lawson ('A') and Captain J. A. Halliday ('B'). The strength of the regiment was:

>Twenty-six officers
>523 N.C.O.s and men
>608 horses.

Haig was not the only soldier of the time to belittle the possibilities, let alone the necessity, of the machine-gun. The 11th Hussars went to war with—precisely two.

These were soon to prove themselves not only useful but vital.

* * * * * *

After detraining at Jeumont in the Maubeuge area on August 18, 1914, the regiment was ready to go into action against the Germans—who still did not know that any British troops had arrived in France, let alone were close to the Belgian frontier—just four days after leaving Aldershot. The B.E.F., under Field-Marshal Sir John French, was on the left of five French armies; the plan was for the French centre to break the German centre 'and cut off the huge German right wing now known to be wheeling through Belgium' (Lumley) while the left-hand part of the Allied force, the French 5th Army and the British Army, was to advance to envelop the Germans' right flank in Belgium.

Thus, as part of Major-General (later Field-Marshal Lord) Allenby's Cavalry Division, the 1st Cavalry Brigade (Brigadier-General C. T. Briggs) headed the advance towards Mons on August 21, the regiment following the leading regiment, the Queen's Bays. The Mons area was safely occupied, and next day the 4th Dragoon Guards had a successful encounter with some German cavalry; so too later did other regiments, but it was clear that the strength of the oncoming Germans was considerable, so the line was held until the infantry (II Corps) arrived to take it over. Then the whole Cavalry Division was switched westwards in an exhausting night march to take over the left British flank.

On the broader scale, however, things were not going well. The French 5th Army had been driven back in heavy fighting between Namur and Charleroi, leaving a widening gap between it and the B.E.F., the British line now being nine miles ahead of the French one. The British advance was therefore cancelled on the eve of the battle of Mons, which started on August 23: worse was to come, for late that night news came that the French 4th Army on the right of the 5th had also been driven back, so that the 5th had had to retire even further. As Lumley records:

> 'This placed the British Army, on its left, in a most perilous position. It was many miles in front of the French line with both flanks exposed. Sir John French therefore decided upon an immediate retirement to a position running through Bavai, some eight miles in rear. This was the beginning of the great retreat, and those were the causes which brought about the difficult days that followed, which were so distasteful and bewildering to the British troops who had already begun to feel that, man for man, they were more than a match for the Germans.'

The history of the retreat from Mons back into France is well known, and the utter confusion that resulted. Part at least of the chaos was the result of the total lack of maps; the 11th, and presumably most other fighting regiments, had been lavishly equipped with maps of the country from Mons to Germany, but had none for a retreat through France. How different, but equally exasperating, was the situation of the leading troops of General Dempsey's 2nd Army in 1944 when because of the speed of the armoured and motorised infantry advance many people ran out of even small-scale maps of vital parts of France and Belgium. Just as in 1914, though more briefly, a Michelin or other guidebook or road map, even a railway timetable book, was worth its weight in gold; one advantage possessed by the advancers in 1944 was their silk 'escape maps'.

The retreat from Mons has similarities to the retreat to Dunkirk. Regiments, battalions, units of all kinds got split up; few knew where they were or why, or where to find their neighbours or headquarters. Supplies went hopelessly adrift, and lack of food and sleep, long, wearisome and often misdirected marches by day and, worse, by night wore out the troops as much as the continuous fighting. 'A' Squadron under Captain Lawson was fired on by our own infantry and cavalry as well as by the Germans near Busigny, and by August 29 Lawson could still count only forty-three men out of 151, though when he rejoined the regiment next day he found another twenty-eight with it. The story was much the same with other squadrons and other regiments during the hectic battle of Le Cateau. On August 26 Colonel Pitman had with him less than half his regiment.

Already the smart cavalrymen and horses of a month earlier were unrecognisable—dirty, tired, footsore. Already they were becoming veterans. Despite all the rumours, disorganisation, shelling, fighting and disintegration, some sort of order was soon restored. But the retreat went on. Much criticised at the time, it was as well that it went on across the Marne, for by one of the imponderables of war a trap was unintentionally prepared for the Germans into which they walked, and of which Joffre took full advantage —fuller than he could have done a little earlier.

One of the most remarkable stories of the war was that of Private Patrick Fowler, who had been cut off during the battle of Le Cateau, and lived through half the winter in woods behind the German lines. In January 1915 he found shelter in the village of Bertry with a French family, and the fantastic tale of how Mme. Belmont-Gobert kept him hidden from the Germans until October 1918 has been told in detail by General Spears in *Liaison 1914*. For most of these years Fowler lived concealed in half of a wardrobe (now in the Imperial War Museum) measuring five and a half feet by five and a half feet by twenty inches. Even more extraordinary, Cpl Herbert Hull, also of 'A' Squadron, who had been cut off at the same time, also finished up hidden in Bertry by a French peasant family, the Cardons. Fowler and Hull met and planned an escape to Holland, but Hull was betrayed to the Germans by a *mauvaise française* (given a life sentence after the war), captured and shamefully treated, then finally shot. Mme. Cardon was given twenty years' hard labour in Germany. The final coincidence came when Fowler, arrested as a spy by British infantry in October 1918 when they liberated Bertry, was being marched down the road for interrogation when the party came abreast of Major F. V. Drake, M.C., commanding the Cavalry Corps Signals, walking from the opposite direction. Fowler, now almost unrecognisable, thin and grey-haired, shouted out, 'That's my troop officer, Mr. Drake!' And so indeed it was, for Drake's troop had been split up near Busigny in the action of 1914. Thus Fowler returned to the regiment. The two devoted Frenchwomen were rewarded (they were very poor) both by the British public and by the regiment after the war, largely through the good offices of Spears and the *Daily Telegraph*. Spears wrote of them:

> 'These women . . . were, uneducated, hard-working peasant women. They were typical of their class. They seemed hard, parsimonious and narrow, but they had hearts of gold.'

Mme. Belmont-Gobert, and her daughter too, received the O.B.E.

By August 31, a scorching day, 1 Cavalry Brigade, still

with its parent division in retreat, had crossed the Oise at Verberie and settled for the night into billets in the small, undistinguished village of Néry, south of Compiègne and its forest and only some fifty miles from Paris. The Bays, 5th D.G.s and 'L' Battery, R.H.A. settled down on the outskirts of the village, the 11th, luckier, were allotted two large farmhouses and the village green. Lawson remarked in his diary:

> 'No news of anything happening outside our own division; we continue to retire and shall soon be at Paris; saw a Brigade of Chasseurs à Pied at Verberie but otherwise none.'

1 Cavalry Division was due to continue its withdrawal, protecting the left flank of the B.E.F., at 4.30 a.m. next morning, September 1. Owing to inadequate and vague orders from G.H.Q. it was not known that the outpost lines of II and III Corps did not link up, and that in fact Néry, which everyone thought to be safe as houses, lay in a large gap between those formations.

The village stands in a commanding position, its church spire a conspicuous landmark, near the top of a tiny stream which threads its way through a deep ravine between two plateaux, one of which commands Néry; the ravine gradually becomes an insignificant valley. In 1914 the farmhouses were surrounded by stout stone walls; in some fields was a beet crop, in others the corn had just been stooked.

September 1 dawned warm and very foggy, with visibility at most 150 yards, sometimes 100. Because of this the brigade, which had saddled up at 4 a.m., was stood down but told to be ready to move at 5.30 a.m. The Bays began to groom their horses, the 11th unsaddled some of theirs and got breakfast, and 'L' Battery, like the Bays camped in the open fields, got their poles down and started to water by sections in the nearby sugar factory above the valley.

On the German side General von Klucks's 1st Army had had little contact with the British after moving southwestwards from Le Cateau; nor had it met much opposition from the French 6th Army on the left of the B.E.F. Von Kluck decided to ignore both enemy armies, turn southeastwards to exploit the 2nd German Army's success at Guise against the French 5th Army on the British right, and possibly finish the war by taking Paris. The reconnaissance towards Paris was the task of Lieut-Colonel von der Marwitz's 2nd Cavalry Corps, protecting von Kluck's left flank, and this Corps was ordered to make for the flank of the French 5th Army in the direction of Soissons. The 4th Cavalry Division, under Lieut-General O. von Garnier, was given the task of reaching the French flank. Thus it had made an all-night march through the forest of Compiègne, and men and horses had had little sleep on the night of August 31, and some formations had got mixed up in the confusion of tracks. But about the time that the British were saddling up the column came clear of the forest, reached the valley of the River Automne just north of Néry, and began the climb out of it. At about 4.15 a.m. or soon after the leading Uhlans reached the top of the plateau east of Néry, where they were relieved by dragoons from another brigade. Shortly afterwards a patrol of the 17th Dragoons came in and reported that a British force was bivouacked at Néry with no suspicion at all of the presence of a large German force. General von Garnier decided to attack at once. The strength of his division was nominally 5,200 men, but it had suffered severe losses at Haelen on August 12 from the Belgians. Despite that and the tiredness of his men, the chance was too good to miss.

For the first fifty minutes of the engagement that followed there were no British troops within three miles of Néry except the 1st Cavalry Brigade—nine squadrons and one battery. Von Garnier mustered twenty-four squadrons and three batteries. But how did the engagement start?

At 4.15 a.m., almost at the very time that the German cavalry were reaching the top of their plateau, 2nd-Lieutenant G. W. A. Tailby, of 'B' Squadron, left on his first patrol with Corporal Parker and five men to reconnoitre the high ground north-east of the village and see if there was any enemy movement in the general area Bethisy St Martin-Bethisy St Pierre one and two miles away. With some difficulty Tailby and his party climbed up the steep slope to the eastern plateau, where visibility was still just as bad on the high ground. After making almost a complete circuit Tailby was about to leave when the fog lifted briefly and he saw, about 150 yards away, a column of cavalry. He knew at once it was German, from the long cloaks and spiked helmets.

The Germans, however, who were dismounted and seemed to have lost their bearings, did not see Tailby. Just then one of his advanced scouts got off his horse and fired at an enemy picquet, not having seen the main body. In Tailby's words: 'That settled it. An order was given and I saw the Uhlans mounting. I shouted out, "Files about. Gallop!"'

Tailby's charger put a foot in a hole galloping down a cart track, refused to get up after somersaulting, and as the Germans were then within fifty yards Tailby plunged into a thicket after shouting to Parker to report back at once. But surprisingly after a few moments the Germans wheeled round and retired, and Corporal Parker reappeared leading Tailby's horse, which had followed his patrol down the track. 'We galloped down the hill, at the bottom of which was an estaminet, outside which was a German cloak and rifle. A woman said that three Germans had just run out of her house. . . . I gave the order to pick up the cloak and galloped on.'

When Tailby reported to Colonel Pitman that there was German cavalry a mile away, the Colonel's first reaction was sharp. He was sure that because of the fog Tailby had mistaken French cavalry, known to be in the area, for German. Exactly the same happened when Corporal Parker warned the 5th Dragoon Guards. The grey-green cavalry cloak clinched the matter, Colonel Pitman sped off to Brigade Headquarters to warn General Briggs, while Major Anderson organised the 11th into defensive positions. Here the farm walls were to be a godsend. Strangely, Major Anderson, who had seen Tailby's patrol off, had woken early with 'a curious feeling that something was going to happen'.

Colonel Pitman had barely returned to the regiment when at 5.40 a.m. the first shell burst among the horses of the Bays, not all of which were yet under cover. It was followed by many others, and prolonged bursts of machine-gun and rifle fire from the eastern plateau and ridges much nearer. There was still shifting fog but the German fire was extremely accurate. The ranges varied from 200–1,000 yards, according to Lieutenant F. G. A. Arkwright of 'A' Squadron, who was brushing his teeth when the barrage started. After finishing saddling up and shutting the horses into small yards, as Arkwright described the scene:

> 'I ran outside with my troop behind me and threw myself down in a gap on the bank by the side of the road. Looking about me, I saw [Major J. S.] Cawley (brigade major), on one side of me with a ghastly wound in his head, obviously done for poor chap, though he was still alive then. On the other side was a gunner, corporal firing away with a rifle quite regardless of bullets all about, and cursing the Germans all the time, saying they had wiped out his battery and he prayed they might all be killed themselves and so on. Just to my right front in a stubble field was the wreckage of L Battery, and a fearful sight it was too. Guns lying all anyhow, a few men crawling about, and bunches of them behind two cornstacks in the field. They got one gun out of the field and the battery sergeant-major and a French officer attached to the Bays kept on with this

at intervals for some time. Lining the bank were my troop, about six of the Bays and their maxim gun, which did great work. Opposite we could see five German guns 800 yards off, but L Battery had silenced three of them before they themselves were snuffed out, and the last gun plus the Maxim gradually silenced the others.'

This was one experienced subaltern's view, commanding some thirty men and horses. Captain 'Deafy' Arkwright was killed in 1915 as an observer in the R.F.C.

How did the small battle look to others?

One of the first German H.E. shells fired came through the roof of Brigade headquarters. General Briggs picked up the time-fuse, saw it was German and set for 800 metres and sent off despatch riders on motor-bicycles to warn Generals Allenby (Cavalry Division) and Snow (4th Division), some three miles away at St Waast and Verberie respectively, of a German attack—strength unknown. Mist was still heavy. Underestimating the size of the German force, pardonably, Briggs gave Lieut-Colonel G. K. Ansell, an outstanding leader, a free hand to use his 5th Dragoon Guards to act mounted against the flank and rear of the German cavalry. The 11th were ordered to take over Ansell's positions as well as to hold their own. Ansell's regiment, like the Bays, had been caught in the open, though not so badly, but scores of horses of both regiments had been killed, wounded, or had stampeded when the heavy German fire started. Ansell led two of his squadrons out of the northern end of Néry to make a wide sweep and left hook. One squadron and his two machine-guns stayed behind to cover his right flank, prolonging the line of the 11th.

Lawson's 'A' Squadron became brigade reserve, and he organised a barricade in the main village street and defensive positions; having met a Bays squadron leader whose squadron had been 'blotted out' he posted two troops to help him, noting 'Bays' maxims doing good work; never ceased firing'. Continuing his investigation very coolly, as usual, Lawson 'found L Battery wiped out except B.S.M., working one gun practically alone. . . . German batteries in action on opposite ridge about 700 yards away.'

The German artillery had been shifted from their first position to the south-east of Néry and had been brought up to that range, or even closer, covered by their own machine-guns with them. This was about 7.20 a.m. They were still on the safe side of the ravine, of course, but all three batteries (and the Guard machine-gun battery) were south-east, or east-south-east, of Néry and within 500 yards of the village's eastern edge. At the same time the German 18th Dragoons advanced, dismounted, against the southern part of the village; one squadron worked round and occupied the sugar factory some 500 yards directly south of Néry, and even got some men across the main road to Rully; the Bays had held up this advance for some time in the factory itself, but were shelled out of it. The dismounted direct German attack in rushes against the village by two other squadrons met deadly fire from the Bays and the 11th and could get no further.

To the north the 2nd Cuirassiers and 9th Uhlans (from another brigade) were just beginning a dismounted attack, with several squadrons still mounted, when Colonel Ansell and his two squadrons loomed up on their right flank. The 5th Dragoon Guards were being led by a sergeant Troop leader, and when the fog lifted briefly he saw over to his right, facing away from him, 'a great, solid tight-packed, immobile mass of German horsemen in profile' (Spears)—presumably the bulk of this German brigade not engaged in the dismounted attack. Without waiting for orders he galloped his Troop to within 100 yards of the Germans, who had not seen him, dismounted them, and opened rapid fire into one of the best targets ever likely to be offered.

The Germans were so densely drawn up that they had no room to turn round quickly or deploy, and they were utterly confused, especially when Colonel Ansell, quickly sizing up the situation, dismounted nearly all of his two squadrons in two different areas and added hugely to the fire-power. The effect on the German attack at this end was devastating. Then the fog came down again preventing a mounted charge, but it had helped Ansell far more than hindered him, because the confused Germans thought his small force much stronger than it was. Colonel Ansell was killed as he directed the fight on horseback, a severe loss, but he had done his work brilliantly, especially by the skilful determination with which he had led his squadrons at speed through fog to the vital area. The sting of the northern flank attack was quite removed.

'L' Battery R.H.A. under Captain E. K. Bradbury had been most hard hit of all when the German attack started. Bradbury had been standing talking to another officer by some haystacks in the field where his guns were bivouacked when the first shell burst over the battery, and then the whole field was swept by intense fire. As the poles were down, the teams had no chance, and when the horses tried to bolt the poles were driven into the ground. Bradbury rallied his men magnificently, and with three of his officers and Sergeant Nelson managed to unlimber three of the guns and get them into action. Three out of eight, with the ammunition wagons twenty yards away. One gun received a direct hit before it could fire at all, the second was soon knocked out, but Bradbury's gun miraculously survived— one against twelve German guns—for well over an hour. Almost the whole detachment was killed or wounded and Bradbury was fatally wounded fetching more ammunition. Finally the few survivors under Battery Sergeant-Major Dorrell and Sergeant Nelson, both themselves wounded, were killed, and they remained firing the gun until there was no more ammunition. The gun was intact.

Bradbury, Dorrell and Nelson were later awarded the V.C. 'L' Battery lost 150 out of its 228 horses, but the job had been done.

It had been done because by the time the last gun ceased firing reinforcements, if they had not physically arrived, were making themselves felt. These were four guns of 'I' Battery, R.H.A. (Captain H. P. Burnyeat) from 4th Cavalry Brigade. Sited some 1,500 yards south-west of the sugar factory, whose chimney they used as an aiming point, they fired very accurately and helped silence the German guns and prevent the German teams from withdrawing eight of the twelve. This was about 8 a.m. Dismounted cavalry from the same brigade worked towards the factory. Soon after the leading company of 1st Battalion The Middlesex Regiment and its machine-gun section, sent post-haste by Major F. G. M. Rowley to help, arrived at the northern end of Néry to join the defence, and later to advance and cross the ravine.

The machine-guns of the 11th under Lieutenant D. McM. Kavanagh had been no less busy than those of the Bays. Starting in a good position by the church they fired very effectively, although early on a mistake by one of the French interpreters cost Kavanagh probably the best target of his life. Thinning fog revealed lines of mounted men on the plateau across the ravine about 400 yards away. Kavanagh was just about to give the order to fire when the interpreter begged him to stop, shouting 'They're French! French Cuirassiers!' Indeed they had looked as if they were wearing cuirasses—if so they were undoubtedly the 2nd German Cuirassiers (Kurassier-Regiment Königin Nr. 2); if not, they were either the 15th or 16th German Hussars. At all events they were not French, though there were supposed to be French cavalry in the area. Later Kavanagh moved his section further south to join the Bays' machine-guns at the bottom end of the village, and between them, with a splendid cross-fire on the German guns, they did most of all, perhaps, to hold up the attack both in the centre and in the south.

General von Garnier was running out of time—and ammunition. He could call upon no suitable reinforcements, whereas he knew that the British would be reinforced very soon, on the ground, and were already backed up by new guns. Further, he had disconcerting reports of enemy to his rear. He tried one last throw, ordering his third brigade to deliver a mounted attack on the southern end of Néry. The 16th and 15th German Hussars charged across 1,000 yards of the plateau but then fell foul of the ravine, having made no reconnaissance. Some of the 16th, trying to ride down a very steep slope, literally fell foul and the charge came to an abrupt end for the 16th. The 15th struck an easier part of the ravine and got to the bottom, where they dismounted and tried to continue their advance on foot with the 18th Dragoons' dismounted line which had already attacked earlier. But the fire from the village remained deadly and the attack ground to a halt.

The Germans had fought bravely but had been undone by the steadiness of their opponents and by the ravine. Von Garnier, in the words of an official German account:

> 'ordered the discontinuance of the battle, and a retreat in an easterly direction. In the meantime the enemy artillery and machine gun fire had caused such heavy losses in the two batteries south-east of Néry that the guns could not be taken out of their positions. The gun crews abandoned their positions, retreating in extended order taking their wounded with them. The enemy did not press the withdrawal.'

The last two sentences present a distorted picture. When General Briggs saw that the Germans were beginning to withdraw, he gave Colonel Pitman permission to send one squadron in pursuit to capture the guns, with strict instructions that it should not go too far. 'C' Squadron was given this job. Major Lockett, unlike the Germans, wisely had reconnoitred the ravine with a dismounted patrol in case his squadron should have to cross it.

He had also sent one Troop to make contact with the 5th Dragoon Guards, of whom nothing had been heard since Ansell's death, and soon an 'all clear' was signalled from their position. At about 7.45 'C' Squadron began to cross the ravine, leading their horses dismounted, about 400 yards north of the German gun position. At the top the leading troop under Lieutenant Willoughby Norrie took a few prisoners and scattered parties of Germans could be seen retiring eastwards. Lockett then ordered 3 Troop to charge the guns while another Troop opened fire on the retreating Germans. Lieut-General Lord Norrie, writing in 1964, describes the result:

> 'We charged the German guns with a rousing cheer and drawn swords, and captured eight guns and two machine guns. This was a comparatively simple task as the machine gun fire, particularly of the Bays, and 'I' Battery R.H.A., had practically silenced the gun battery. For myself, it was a great thrill to lead a cavalry charge, even if the enemy surrendered shouting "Kamerad, Kamerad!" and put up little resistance. My Troop Sergeant (Sgt Hailey) stuck one who was too slow in putting up his hands. We then pushed on to another village [Le Plessis Chatelain] where we captured a further 100 prisoners until we were ordered back, and what a terrible sight it was to see those gallant men of 'L' Battery, surrounded by so many dead horses. . . .
>
> 'The conduct of 'L' Battery and the Bays who were caught in the worst positions was outstanding. . . . We must give full credit to the German 4th Cavalry Division, who had marched 400 miles and fought two major battles. Without any hesitation, they attacked. There was no lack of dash by the German cavalry, even if there was lack of reconnaissance, but how lucky we in the 1st Cavalry Division were to have that ravine between us and the enemy.'

These German guns were the first to be captured by the B.E.F. in the war. In his first charge Norrie had taken about twenty prisoners, and then had wheeled about and captured a few more runners. A Corporal and six men of the Middlesex Regiment then appeared from the ravine—the advanced patrol of 'D' Company—who had been sniping from 150 yards at the guns, five of which were pointing towards 'L' Battery, three towards 'I' Battery. The guns were handed over to the Middlesex, and about half an hour after Lieutenant Norrie had galloped the guns more companies of that fine regiment arrived on the scene. It was with these that Major Lockett arranged the capture of Le Plessis Chatelain.

The fight at Néry was over. It had been spectacular, and a fine example of co-operation between cavalry, their own gunners, and all arms of the 1st Cavalry Brigade—and, not least, an example of rapid and sensible reinforcement and help by 4th Cavalry Brigade and by 4th Division. The Middlesex were easily the first infantry on the scene, but others had rushed to help only to arrive when the fighting was almost over. As Lord Norrie recalled half a century later, Brigadier-General Briggs 'set a splendid example of coolness', and everyone had complete confidence in him. This was by no means always the case with the troops and the 'brass hats' later on. General Sir Evelyn Wood, V.C., who had fought in the trenches in the Crimea as a Midshipman, would surely agree that Néry provided an excellent example of what he considered the essential quality in cavalry officers—that they 'should know when and how to charge, and when to refrain from the attack'.

The four guns that the Germans had managed to get away were found abandoned in a wood next day twelve miles south of Néry, for General von Garnier, surrounded by British columns, had boldly stuck to his original destination and made for Rozières—which he reached—the Marne and Paris—which he did not. The division was split up at Rozières into three columns. If they had not run out of ammunition during their wanderings they might have played havoc, as Lumley says. 'As it was they frightened G.H.Q. out of its billets at Dammartin and exchanged shots with the escort of III Corps Headquarters.' But the division was no longer a division and everyone was exhausted.

This shows the results of Néry, and of other encounters. Von Garnier's division did not continue the advance with von der Marwitz's 2nd Cavalry Corps, but was left behind in a Reserve Corps doing right flank guard to von Kluck's Army. Because of its losses, particularly in guns, and its exhaustion, it was kept much too far back to do proper

Imperial War Museum
11th Hussars on the march in France 1916.

reconnaissance, and as a result von Gronau's Corps blundered without warning into General Maunoury's Army on September 5. An active cavalry division could have warned von Kluck what he was coming up against, but it took von Gronau some time to disentangle the situation.

This failure to discover in time the threat that menaced von Kluck's open right on September 5 as he rushed southwards was what really led to the ruin of all German hopes on the Marne. It is a failure that can be traced directly to Néry.

The 11th Hussars had been amazingly lucky. They had lost only two men wounded, and two horses, when other regiments had suffered severely. It was sheer luck that they had been among stout farm buildings and not in the open. But it was not luck that had made Colonel Pitman reject any temptation to put the horses in fields for the night, and insist on the farms being occupied. Major Anderson's apprehension about Tailby's patrol might be called luck, but he acted upon it and ordered the squadrons to rehearse their defensive positions briefly.

Perhaps the last word should go to the Germans, indeed to von Garnier himself. An intercepted wireless message after the Néry battle read:

> 'Have been attacked by the English at dawn and am unable to fulfil mission.'

Chapter 10

The Great War: II

'At this time—it was at the end of 1914—the French, who considered themselves a military nation, had not yet realised that a dark blue coat and red trousers for infantry, a cuirass and helmet with plume for cavalry, and soup squares for rations in a water-logged trench were unsuitable. They still had much to learn: so had we.'

(Major Robert Hartman, *The Remainder Biscuit*, André Deutsch, 1964, p. 96.)

ALL retreats come to an end, and the Battle of the Marne, which saved Paris and decisively defeated the Germans, was followed by the advance to the Aisne, and then the deadlock there. The regiment took part in all this fighting, but from mid-September onwards had its first taste of trench warfare, as the Germans tried desperately to drive the B.E.F. back across the Aisne. The deadlock in this part of France, instead of being broken, gradually was repeated up the line, and as neither side could turn the other's flank barriers of strong positions and trenches were extended, not so slowly, which before long reached the sea. The deadlock, in fact, was not to be broken for four years.

When it became obvious that no further advance could be made north of the Aisne, Sir John French began to switch the B.E.F. back to its original position on the left of the French armies, with Joffre's agreement; thus the Cavalry Corps, after a week's tedious day and night marches, found itself in the extreme northern corner of

France, in the Somme, Pas de Calais, and Nord departments, near the Belgian frontier—in the Lille coalfield area, in fact, briefly so well known by 11th Hussars thirty years later. Soon names like Ploegsteert, Menin, Ypres, Messines, Armentières, Nieppe (or their vernacular equivalents) were to become trenchhold ones.

In the advance from Béthune to the Lys River the 11th distinguished itself, often leading the brigade or division, and in securing the Lys crossings it played its part. Its reconnaissance in flat enclosed country dotted with houses and farms (always difficult) had meant casualties—all of which occurred mounted, at close range—but the necessary information had been secured. Of this period between October 11–20 Sir John French wrote in his book *1914:* 'Of all the splendid work performed by the cavalry during the War, little can compare in results achieved with this advance.' A large tract of country had been cleared of enemy, and the cavalry regiments had performed all the tasks required of them in an advance. But now the race to the sea was over, with neither side the winner, and over, too, unfortunately, was mounted fighting cavalry work for the regiments of 1914.

The Flanders battle of 1st Ypres was the greatest achievement of the original B.E.F., heavy though its cost was to the Regular Army. There was bitter fighting for many days, and many units were reduced to remnants—but had the Germans not been stopped they would surely have gone on to win the war. As it was, the British were not attacked again with an all-out decisive effort for more than three years, and precious time was won for Kitchener's new armies, and those of the Empire, to be raised and trained. As often happens, the B.E.F. was committed to an attack itself when the Germans struck, having launched an entirely new Army, the 4th, of four Corps mostly consisting of young volunteers between seventeen and twenty with some reservists. This Army had been brought up between

Menin and the sea entirely unreported by our Intelligence. It attacked the Belgians on October 18 and 19, but they stood firm. On October 19 the advancing Germans collided with General Rawlinson's advancing British from IV Corps and next day the first attack on the British line was made. Thereafter the untried but desperately keen 4th Army and the already seasoned 6th German Army to the south made attack after attack for three weeks on the British line, first east of Ypres and then south-east of it, so that gradually almost the whole of the B.E.F. was called upon to stem the torrent. Lumley writes:

> 'The part played by the cavalry in this battle was in no way different, either in importance or in the manner of fighting, to that played by their infantry comrades. They fought as infantry in the line wherever they found themselves when the battle began, and several of the villages round Ypres, which have now become historic, were first made so by British cavalry regiments. For the Eleventh Hussars the village of Messines became the scene of perhaps their most important work in all their history.'

The regiment, having advanced to the 'Plugstreet' area, met, and then 'seen off' the Germans between October 19–23, was then sent to the Messines area when Allenby decided to shorten his line so that he could form a reserve and handed over positions south of the Douve stream to III Corps. But on October 21 the two brigades of 1st Cavalry Division had easily dealt with an attempted advance by three German cavalry divisions between Ploegsteert Wood and Messines. The superior training, and especially accuracy of fire, of the British made this possible.

Messines was a small Belgian village, but then as now its position was important; standing on the southern end of the ridge which runs some two miles south-eastwards from Wystschaete—the famous 'Messines Ridge' as it came to be called—it dominates the flat surrounding country, offering excellent observation for miles except on the Wulvergehem Ridge side to the west. It is 'a formidable barrier to any army that wishes to pass south of Ypres either from the east or from the west'. (Lumley.) In 1914–18 it was shelled heavily and frequently.

Between October 22–30 the 11th were in and out of the front-line trenches—the detailed communicating trench system of following years was not yet in being, and there were no continuous lines of trenches in depth yet—alternating with the Bays, the 9th Lancers, and the 5th Dragoon Guards, and usually with the Connaught Rangers or the 57th (Wilde's) Rifles on one flank or the other. The latter suffered the first Indian casualties of the war: both regiments had been brought up by London buses to help the cavalry at Messines, reminiscent of the surge of French taxi-cabs at the time of the Marne. The splendid shooting of the Connaught Rangers did not outmatch that of the 11th, and, as was frequently the case with both infantry and cavalry regiments early on, the Germans often mistook the rapid rate of fire from the new Mark III Lee-Enfields, expertly handled, for machine-gun fire.

The Germans too had dug in. Their trenches also were manned by cavalry as well as by infantry, and the former on one occasion obliged 'snipers' of the 11th by signalling back the results with a spade! On October 25, Balaclava Day, Colonel Pitman noted the fact, adding: 'and the 11th Hussars were digging like moles under ground.' 'Very wet night, the trenches liquid clay . . . every twig and bush was a German in a gale of wind, mud, rain.' The next day, when he was relieved by the 5th Dragoon Guards and the regiment went back to Messines, he observed of them: 'a sorry sight, smothered from head to foot in liquid clay, and the rifles completely clogged.' As the 11th marched into Messines,

> 'The German shells had set alight to the church, which was now all ablaze. A golden spire stood up one

mass of flames, but soon gave way, and the beautiful building, 1,100 years old, was completely gutted. Against the blaze an angel, suspended by a chain, was silhouetted, quite unscathed, as if the fire could do it no harm. The following morning all that remained were the four walls and a wooden crucifix which the flames had not touched.'

The pressure was now on at 1st Ypres. Far from relaxing, the Germans brought in another six divisions and made an all-out assault. On October 30 all three squadrons of the

Imperial War Museum

A machine-gun detachment of the 11th strip their weapon in a quiet moment in the trenches.

11th were in the front line of trenches, and Messines was clearly among the first targets. The British front was attacked from the Lys to Zonnebeke, where it could muster only eighteen heavy but obsolescent artillery guns against 260 heavy German guns. The task of the six new German divisions was to break down the British line from Messines to Gheluvelt and to break through to the Kemmel heights. If they did so they would cut off all Allied troops in the Ypres salient.

On the Messines front, the thin white line of British cavalry had three divisions—equivalent to less than a brigade of German infantry in numbers—with two infantry battalions; their artillery was seven batteries of horse artillery and two 6-inch howitzers. Their front was nine miles long. Their opponents were three fresh infantry divisions—six brigades—supported by a huge mass of artillery. The odds have been estimated at six to one in men and far more in guns. A captured German order of the day for October 30 said confidently, 'the break through will be of decisive importance.'

Equally decisive, and recognised as such by the German General Staff, was the capture of the Messines-Wytschaete ridge, against which the main effort was directed on October 31; for on the previous day, despite very heavy shelling, and a strong attack against Haig's magnificent I Corps away to the north-east, the day had been 'quiet' at Messines. But not so quiet. A heavy bombardment of all the trenches and of Messines started about 8 a.m. on October 30 and lasted through the forenoon. At 10.40

> 'Halliday staggered into Headquarters shelter, seemed quite knocked out, said that 'B' Squadron trench had been blown in, and most of the squadron buried. Four coal-boxes had landed along the front of the trench, which had been considerably undermined in the endeavour to get shelter from the rain; one troop and a half had been buried beyond all hope of recovery.'

[Captain Halliday was commanding 'B' Squadron; he had joined in 1898, and was fatally wounded on October 31; Sergeant T. Frane, who had rallied panicking men when their trench was blown in, was awarded a D.C.M. 'Coalboxes' was the nickname for 8-inch howitzer shells, because of the black burst of smoke; sometimes they were known as 'Jack Johnsons' or as 'Black Marias'.]

Later the shelling died down, and German attacks were put in, but driven off. 'A' and 'C' Squadrons stuck to their trenches throughout the day, and 'B', who had had the hardest day, were driven out from time to time but finally reoccupied what parts of their trenches were not demolished. At 5 p.m. the 11th handed over to the 9th Lancers and withdrew into Messines to take up the inner line of defence there.

There were two cavalry brigades defending the village, with infantry. The 9th Lancers and the 4th Dragoon Guards came from 2nd Cavalry Brigade, the 18th Hussars had been detached from that brigade for the time being, but the Bays, the 5th Dragoon Guards and the 11th were in Messines as the normal troops of 1st Cavalry Brigade.

The main German effort was against Messines and Wytschaete, against which yet another German division was added. The number of German infantry battalions directed against Messines itself was twelve, and the Kaiser arrived during October 31 to boost morale from a safe distance behind the attack.

The 2nd Bn. Royal Inniskilling Fusiliers was relieving the 57th Rifles (in turn having relieved the Connaught Rangers) when the first big attack began on October 31 about 4.30 a.m. An earlier German attack between 1–2 a.m. had been repulsed easily.

About 4.30 a.m., while it was still dark, the enemy attacked again, but were beaten off by the Bays and the 9th Lancers. But they had a local success against the 57th Rifles, and kept the front line of trenches until counterattacked by the 57th, the 5th Dragoon Guards and 'C' Squadron, 11th Hussars under Captain M. L. Lakin and the machine-gun section. This counter-attack more or less nullified the German gain.

Soon after daylight the German bombardment became very heavy. The 9th Lancers were enfiladed under that cover and had to retire into the rear end of Messines, after many casualties. They reformed there, and 1st Brigade took up the defence of the village, with 4th Dragoon Guards replacing 9th Lancers.

German shelling became even heavier, and house after house was smashed, as infantry came on following up in great numbers. An 11th machine-gun firing from a top window of one of the few houses that remained standing did excellent work, as did Captain Lakin's squadron.

About 10 a.m. the 4th Dragoon Guards received orders to retire; as a result the 5th Dragoon Guards could hold on no longer and had to withdraw to the line of the main street. 'B' Squadron, 11th, was sheltering behind a convent wall, but that was also blown down, and 'C', finding its left exposed, also withdrew to the main street. This was probably the worst place of all for shelling. It seems so, because about 10.30 a.m. Colonel Pitman, Major Anderson, Major Moberly, and the Adjutant, were all wounded there. Colonel Pitman remained in the chair— literally a chair, for he directed the battle as has been said 'like Marshal Massena at Wagram'—until the infantry came up to relieve the situation, and he handed over command to Captain Lawson. But before that much had happened.

The German attack went on. The line of the main street was held, but half the village was in German hands. It was now very much an infantry battle, and accounts, times, even dates vary. What is certain is that the Germans were beaten back.

The enemy now held half the town, but could not make

further headway against the defence, which called on every man to act as sharpshooter, infantryman, and dismounted cavalryman—with the bayonet it necessary. There was hand to hand fighting in the streets, house by house being disputed. Pitman relates: 'The Germans attempted to rush across the square, but were shot down by men posted in the opposite windows.'

A German infantryman at Messines wrote:

> 'Here a terrible struggle ensued [in Messines], in which the English fought in a way to compel our highest respect, at least as regards their soldierly qualities.
> 'In the village itself they had occupied every house, the streets were barred by barricades—metres in thickness, the windows of the houses stuffed up with sandbags, the doors barricaded with furniture and stones; through the walls they had drilled loopholes, scarcely visible; each house was a small fortress scattering death and destruction.'

The arrival about 3 p.m. of the 2nd Bn. K.O.Y.L.I. and the 2nd Bn. K.O.S.B., relieved the pressure, but just about that time the British supporting artillery, believing that the cavalry had been beaten out of Messines, shelled the village heavily.

All in all, Messines was a great victory—but a heavy one. Every cavalry regiment paid its price. The total casualties between October 22–November 10 being exactly 100. The strength of the regiment in the trenches was never more than 220 all ranks.

In the end, the Germans captured Messines on November 1, but the losses sustained by them during the previous two days were shattering, and a great deal of wind had been taken out of the full-scale German attack.

In Pitman's opinion, the bad days between October 30 and November 10, 'were the most critical days of the war'.

This seems accurate, although in every account one reads

of the battle there are discrepancies. This shows how severe, tiring and bitter it must have been, and no one is to blame for conflicting accounts. In the end, it all adds up to the same thing. The Germans were stopped, and Messines, a small consolation prize, was to look very different in 1917.

During April 22-23, 1915, the Germans initiated a new and horrible form of warfare, the use of poison gas, of which the first victims were two French divisions holding

Imperial War Museum
The 11th Hussars moving along a canal in France, 1917.

the northern part of the Ypres salient. They were utterly unprepared for gas—as were many British and Canadian troops not long afterwards—and the Germans 'walked in' to the gap of four miles they created. Fortunately they did not make a mass attack after the first gas attack but advanced two and a half miles and dug in. This set the pattern for the battle known as 2nd Ypres, a long and expensive one. The Germans had great superiority in artillery, would attack in sharp rushes after heavy bombardment, and dig in again. To everyone's relief on the British side the War Office reacted to the use of gas with extreme speed and efficiency, so that by early May nearly all troops had been issued with the first type of gas mask—but not before thousands had died or suffered irreparable damage to their lungs.

One or two incidents may suffice to show the regiment's part in this new battle. On May 13, a day in which it rained constantly and the trenches were full of mud and water, the Germans launched their heaviest bombardment so far of the war. It affected nearly all the British cavalry, having lasted over fourteen hours from 3.30 a.m. to 6 p.m. in one place or another. Regiments from six cavalry brigades were affected, and the official histories of all tell the story of a grim, eddying day of battle.

Brigades and regiments all helped one another, and until the day was won most were split up or diverted from their original formation for some time. This fighting was just east of Ypres, and the 11th were in reserve trenches in front of Potijze Château—in the grounds of which 1st Cavalry Brigade, temporarily under Colonel Pitman, had its headquarters. Major Anderson was thus commanding the regiment. The Bays and 5th Dragoon Guards were in the front-line trenches some 600 yards further forward when the day began, and bore the brunt of the shelling here, both suffering heavily. One odd effect that the German bombardment had on Potijze Wood was to start a chorus of protest from dozens of nightingales which went on singing throughout the day.

The Bays held their line magnificently all day, and were helped first by two Troops of 'C' Squadron, guided by Lieutenant Renton, delivering ammunition to them, from 8.30 onwards, and from 10.30 by two Troops under Lieutenant Norrie actually reinforcing them and the 5th D.G.s in the line.

For some time the right flank of the Bays was unprotected, as the Germans had broken through the next-door brigade, but at 2.30 p.m. a very gallant counter-attack from several more cavalry regiments across open ground restored that situation. The 10th Hussars played a fine part here.

It was just at this time that the full force of the bombardment was turned on 1st Cavalry Brigade, and this time the 5th D.G.s, who had a large part of their trenches blown in, were unable to hold any longer, and most were ordered back to the reserve trenches. Lieutenant Norrie's two Troops on their right stayed firm, and later Colonel Pitman ordered Major Anderson to retake the lost trenches—for the situation to his left, where the 18th Hussars had also suffered very severely and been driven out of their trenches, was causing alarm too, and it looked as if the Germans might exploit both successes.

'A' Squadron was given the job. Captain Lawson, having watched the additional heavy losses suffered by the 5th D.G.s as they were caught retiring over open ground, used a slight depression in the terrain in his advance. His description reflects his usual coolness and seeming unawareness of having done anything unusual:

> 'got troop leaders and told them follow by troops and went ahead myself to pick best way: crawled up wet and dirty ditches and got to trench, where found about squadron of 19th, troop of Norrie's, troop of 5th Dragoon Guards and about thirty 1st Life Guards; put all mine in on left to fill up gap between us and 2nd Brigade; got all

in with only two men being hit, though how escaped so lightly don't know. Took command of the mixed force and set about trying to repair trench; still being shelled so couldn't do much.'

Very effective machine-gun fire from the 9th Lancers on the 2nd Brigade side prevented the Germans penetrating far there. The line was now held until the bombardment ended and dark came. Lieutenant R. A. L. Hartman, who had already distinguished himself with his Troop in the van of 'A' Squadron's advance, added to his good work by going out after dark with Corporal Morey to identify some heads moving above a parapet in a trench to the extreme left rear of this little mixed force. This was a nasty job as it was thought more than likely that their owners were Germans. Happily they turned out to be some of the 4th Dragoon Guards from 2nd Brigade. Hartman was to serve longer more continuously with the regiment than any other officer, ending the war in command of his squadron, 'A'; despite some hair-raising experiences, he was never wounded (although hit). On this day he was blown by a shell-burst out of a ditch along which he was crawling, the two men behind him being hit.

But 2nd Ypres was not over. Hooge, the Menin Road, Railway Wood, Zouave Wood, and Sanctuary Wood were to become ominous names. The regiment had got to know them well, but May 24, 1915, stands out as 'one of our worst days during the war,' as General Pitman wrote speaking generally. Very early the Germans let loose another fierce bombardment, accompanied this time by the largest discharge of poison gas that they had yet used. This reached a height of forty feet in dense clouds, and later on the effects were felt as far behind our lines as twenty miles. Respirators had been issued but they were still by no means perfect, and many of the 5th Bn the Durham Light Infantry and some of the 9th Lancers close to the 11th in the Menin

Road–Hooge area were badly gassed. So too were other regiments. Major Hartman writes:

> 'The gas had passed in front of the 9th Lancers, over the Durhams, and behind the XI Hussars. If ever the regiment had a stroke of luck this was it.'

As brigade communications once again were soon severed, Colonel Pitman commanded the 1st Brigade line for most of the day. It was a day of bitter, close fighting with trenches being taken, partly taken, recaptured, and the Germans were much better supplied with machine-guns and 'bombs' or grenades than the British. They gained a considerable amount of ground north of Hooge and got across the Menin Road, but the fighting swayed back and forth and eventually they were driven back across the Menin Road; all their gains could not be crossed off, but Hooge was held, mainly due to the fine defence of it by a squadron of 9th Lancers commanded by Captain Francis Grenfell, V.C., who was killed there, the remnants of the D.L.I., and the 11th Hussars. The Germans captured part of one vital trench and fired and bombed down it all day, blowing up a series of hastily built barricades. At one stage Lance-Corporal H. J. Skipper and Captain Lawson were not only throwing their own bombs but several times caught German grenades in the air and threw them back. Eventually with the help of the Bays and part of the East Yorkshire Regiment, and with a successful attack by the York and Lancaster Regiment, the situation round Hooge was stabilised. It had been a costly day, but a successful one, for it marked the end of 2nd Ypres.

Once again the 11th had been amazingly lucky in casualties. On May 13 they had missed the worst of the shelling, and on May 24 the gas—by fifty yards. Their casualties totalled three officers killed, seven wounded, twenty-four other ranks killed and thirty-three wounded—about the lowest in the Division, with those of the Bays.

The next two and a half years were extremely frustrating for all the cavalry. They were hamstrung operationally for two reasons: first, in semi-static warfare they were no longer needed as first-line reserves to the infantry, because of the huge expansion of the surviving B.E.F. by New Army and Territorial divisions; and second, they were kept by what now seems blind wishful thinking trained and ready as real cavalry for a break-through that never came.

This is not to say that there were not periods of extreme danger, collectively and individually. There was always the risk of shelling, sometimes bombing or strafing from German aeroplanes, even disease, quite apart from the occupational hazards of the periods of trench life, from snipers, to machine-guns to pneumonia in rain, mud, and cold. But in general from June 1915 to November 1917, as Lumley writes, life was confined 'to the uninspiring round of long periods in billets, occasional gaps, now and then a turn in the trenches, and frequent digging parties'. To which may be added training, both mounted and dismounted, and seemingly incessant marches and countermarches. The tedium and pointlessness of this sort of life after a time would make shelling or action welcome.

On September 8, 1915, two extra machine-guns were issued. The regiment now had four in all. It was to be July 1917 before each squadron possessed two Hotchkiss guns. By July 1916 complete mastery of the air had been achieved by the Allies—as far as the ground troops were concerned; and the same month the first news reached them of an exciting new weapon—the tank. Here was a chance for a cavalry break-through to be caused. The soldiers so disappointed in the 11th, as in other regiments, can little have guessed after the breakthrough in September failed to materialise (through no fault of the tanks, but perhaps because of their premature and unskilful deployment) how closely they were to be associated with these strange 'land battleships' in years to come.

Cambrai, which started on November 20, 1917, was a different story, and the tanks achieved with great infantry help what might have been a decisive break-through, the cavalry waiting to exploit it. But wrong formations in the wrong places and breakdowns in communications allowed the Germans to regroup and reinforce—always fatal—and the battle, in which the regiment played a small but useful part, closing a gap near Anneux with 'B' and 'A' Squadrons under Captains White and Yates, lingered on until December 7. It was to be followed in March 1918 by the last great German bid for victory.

The 11th, since 1915 and Colonel Pitman's promotion commanded by Lieut-Colonel Anderson, were still in 1st Cavalry Brigade and Division, and were thus part of Gough's 5th Army. Between March 21 and April 4, when the offensive finally collapsed, they fought at least three splendid small actions, helping Gough save the day.

On March 21 at Vermand near St Quentin Colonel Anderson had been put in command of all troops of 72nd Brigade south of the Omignon River as far as the brigade's west boundary, and had handed over the regiment to Captain Honble Luke White. Anderson was commanding over nine companies from different infantry regiments; his own regiment, dismounted, consisted of twelve officers and slightly over 200 men. On March 22 the Germans started a series of very heavy attacks, the first through fog, coupled with extremely heavy if inaccurate shelling of Vermand. The British positions were covered by barbed wire and the Germans suffered terrible losses from the rifles and machine-guns of the 11th and the infantry, hundreds being counted lying dead in the wire and hundreds more beyond it, all round Villecholles just east of Vermand. Broadly speaking, 'C' and 'B' Squadrons were fighting alongside companies from the 9th East Surrey Regiment and the 12th Sherwood Foresters, with 'A' next to the 138 h Bn Argyll and Sutherland Highlanders, but units were divided up and sometimes mixed up as the battle continued. The Germans managed to get machine-guns into enfilading positions on one flank and into a commanding position from which to fire into Vermand on the other, and with all wire communications cut once again the situation was thought further back to be worse than it was. A resulting British artillery barrage, more accurate than the original German one, did not help and inflicted casualties on its own troops.

At 11.40 a.m., therefore, 100 minutes after the first attack, Colonel Anderson was ordered by 72nd Brigade to withdraw fighting through the reserve 'Green Line' held by 50th Division to reform, because British troops north of the Omignon had already fallen back and the left flank of his force was seriously endangered. Orderly withdrawal in these sort of circumstances is easier said than done, and Colonel Anerson's first problem, with no contact with his units and no runners left, was to get the order through. Lieutenant L. R. Lumley, who had just this moment managed to rejoin the regiment illegally from a musketry course, volunteered and took four men to find the C.O.s of the Sherwood Foresters and East Surreys, and of course Captain White. He also had to see at all costs that the river bridge between Vermand and Villecholles was blown up.

Lumley's first encounter was with the Colonel of a Middlesex battalion in the grounds of Vermand Abbey, who told him where to find the Foresters and undertook to get the bridge blown. Next he reached a company commander of the Foresters, not being able to find the Colonel, and it was agreed that they would begin to retire in ten minutes. Continuing his hazardous journey he stumbled by luck on a dug-out in Mont Huette Wood which contained both the C.O. of the East Surreys and Captain White. In Lumley's words:

> 'They were both distinctly annoyed at the order to retire. The enemy barrage had gone over their line, and they had easily beaten off two attacks with heavy

casualties to the Germans; and so they were quite confident of holding their line.'

Lumley had to explain the situation. He then went on even further to rejoin his own squadron with the Argylls, being hit by a German bullet en route, and being ordered by White to get back to Vermand.

Meanwhile the Germans had not been idle, and were closing in rapidly on both flanks and waiting to pounce in the middle. Thus as soon as the withdrawal began the Germans rushed forwards with machine-guns. German aeroplanes, flying very low, swooped in to bomb and strafe and a panic developed among some units as hundreds of men streamed back in hopeless disorder irretrievably mixed up—including some from other brigades and even divisions. The East Surrey C.O., Major Clarke, had already seen this danger and he and Captain White somewhat restored the situation by making a stand. Captain White walked up and down a trench three feet deep, which was being enfiladed, rallying men of the 11th to turn and fire, with absolute coolness.

But in general there was hopeless confusion, as German artillery joined in and groups of their infantry could be seen looming up to turn the retreat into a rout. Complete disaster might have occurred when the hoarse voice of Major R. G. Moir, M.C., was heard shouting 'Stand fast: the Argylls!' Other officers from different regiments took up similar cries and the situation was saved as groups of men reformed round their officers, took extended order, lay down and fired back at the pursuing enemy, even when shells burst among them in the open. The near-rout became a proper withdrawal.

Lumley's verdict on this astonishing sight, of which he had a grandstand view after his wound, was:

'It seemed to me that if men could recover their steadiness so soon after what looked like the beginning of a panic, the issue of the battle would be all right as long as there remained sufficient leaders of the kind of that Major in the Argylls.'

Hartman, whose squadron had been sharing Spooner redoubt at Villecholles with Major Moir's battalion, described him as 'one of the most splendid soldiers that it was ever my good fortune to see in action'. But typically Lumley, in an account written years later, gave no credit to his own achievement, which had in fact saved everyone in Spooner Redoubt and many others elsewhere.

After a successful holding action at Montauban, near Albert, on March 25, on the extreme left of 5th Army's front, the 1st Cavalry Division, and indeed all troops north of the Somme, were ordered to come under command of 3rd Army. This was because a gap of some three miles between the two British Armies had been forced by the Germans, or in part developed naturally, in the Montauban–Delville Wood area, where the regiment fought that day. The pattern was that gaps were continually being forced along the line, and as one was plugged another developed. The result of this transfer of troops from 5th to 3rd Army, as might have been foreseen, was just that: the gap shifted again, becoming wider in the process, to a new and more dangerous area—along the Somme itself. Amiens was now in great danger, and on the night of March 26 3rd Army ordered that the ground between the Somme and the Ancre must be held at all costs. On March 27 1st Cavalry Division carried out this order, and there was hard fighting round Sailly Laurette on the northern bank of the Somme; unfortunately the Germans poured troops across the river behind 5th Army through part of the new gap and their way south of the river to Amiens looked open. 1st Cavalry Division was hurriedly withdrawn across the Somme that night and for the final week of the battle held the line near Hamel—just opposite Sailly Laurette—and helped block the way to Amiens from that approach.

About 4 p.m. in the evening of March 27 the 4th and 5th Dragoon Guards withdrew from Sailly Laurette under great pressure on three sides; the 11th, which had originally been with them, had already been sent back to the nearby village of Sailly-le-Sec to meet an emergency on either side of the river. A 3rd Army order had arrived just after 3.30, but had not been in time because of bad communications, the confusion of the fighting and its fluidity, directing present positions to be held at all costs.

Colonel Anderson was therefore ordered to return to the village and rejoin the Dragoon Guards—who had of course left already, unknown to Brigade, which was confident that they were still there. Anderson decided to make his advance dismounted from the south, which was as well, for it soon became clear that Sailly Laurette was held by the Germans. Colonel Anderson decided to attack immediately with the bayonet—by this time he had only about 120 men available, for it was the seventh day of the battle. Harassed by machine-gun and rifle fire from the south side of the river the 11th attacked, led by Lieutenant H. A. Jaffray of 'A' Squadron, who ran round a house straight into a party of unsuspecting Germans, shot the first, but was thrown to the ground by the others. His squadron leader, Major Claude Rome, and another party appeared suddenly just in time to save him, Rome, who won two D.S.O.s in the war, 'roaring like a bull' and firing his revolver. The Germans surrendered. The rest of the regiment then charged with the bayonet, completely surprising the main body of Germans in the village, who had only just arrived. In a few minutes they had retaken Sailly Laurette. Further German attempts to retake the village, including one from the 'wrong' side of the Somme, were beaten off, R.S.M. Upton knocking off the first four Germans who tried to cross by a rickety bridge from the south, *pour encourager les autres*. He was, after all, a crack shot.

Until the whole division was ordered to retire later in the

day to new positions the 11th held the village. Their counter-attack, completely successful, had been a pleasant contrast to the continual retreats of the previous week. A 3rd Army staff officer who watched it reported it in glowing terms, 'with the result that, much to its surprise, congratulatory telegrams poured in on the Regiment during the next few days from London and Paris.'

Finally, on March 30, south of the Somme now but only a mile or two away, the regiment took part in the defence of Hamel against very fierce German infantry attacks. These were preceded by what many thought the worst bombardment by artillery and mortars that the 11th had endured during the whole war. And when that stopped hedge-hopping German planes machine-gunned the defensive positions and trenches. But the Bays and the 11th—the latter with a scratch force of several hundred under command, ranging from 200 Royal Engineers to 150 men of the United States Corps of Engineers, and including many stragglers—held out; the 5th Dragoon Guards recaptured a vital trench with the bayonet and by evening the whole of the 1st Cavalry front was stabilised. The defence of Hamel had saved Amiens.

During the period March 21–April 4 the 11th lost three officers killed, ten wounded, one taken prisoner; twenty-four other ranks killed, seventy-six wounded, nine taken prisoner. Compared with infantry losses these figures may seem small; not so small when one remembers the strength of the regiment. All the cavalry were constantly being withdrawn from their battle positions and sent without rest to plug dangerous gaps.

With the end of the last great German throw the end of the war, although not yet in sight, was not everlastingly distant as often it had seemed until now. After August 8, the 'Black Day of the German Army', it became certain.

The 11th took part in the final advances, and on November 11, 1918, found itself in the small village of Lens, in

Belgium, near Mons—not so far from where it had first seen action in 1914.

It was here, not long after the horses had been watered and fed, that Private 'Algy' Knight, a future immortal 11th Hussar, cycled into the village and delivered the following message to the Adjutant, Captain David Margesson, M.C.:

> 'Hostilities cease w.e.f. 1100 hrs today 11th November. No further movement eastwards. Fraternising with enemy not allowed. Ack or words to that effect.'

At last it was over. As always, immediate relief took time to sink in properly.

The 11th, like other cavalry regiments, had borne an easy lot compared with that of the infantry, British, French, and American. For most of their infantry comrades they had come to have the highest admiration—the more so because they had sometimes done the same job. The regiment had lost from its own ranks 163 killed, 337 wounded, and sixty prisoners—casualties greater than the numbers it had started out with; but it had also lost many other former 11th Hussars dead and wounded elsewhere with other regiments.

Chapter

11

1919–1941(1)

'Off we go, into the wild blue yonder. . . .'
(Apologies to the United States Army Air Corps and
the United States Air Force.)

THE 11th Hussars were among the first British troops to enter Germany on December 1, 1918, and exactly a week later they moved up to the Rhine and were billeted at Bad Godesberg. Almost twenty years later this name must have rung a suspicious-sounding bell in the minds of many who had been there in 1918 as Chamberlain returned from his second conference with Hitler; even more so after Munich and 'Peace in our time.'

The Rhine was crossed at Bonn on December 12, but the stay east of it was brief and three days later the regiment returned to the Cologne area where it remained until its return to England in March 1919. The next few months at Aldershot were spent building up an almost completely new regiment because of demobilisation and retirements, and when the 11th sailed for Egypt in July their ranks were filled almost entirely by young recruits.

In 1920 Colonel Anderson, C.M.G., D.S.O., gave up command after twenty-seven years with the regiment; he had been with it throughout the Great War, in command for most of the time. Lieut-Colonel W. J. Lockett, D.S.O., succeeded him. Also this year H.R.H. the Duke of York (who had made his first visit to the regiment in Germany) was gazetted Colonel-in-Chief of the 11th Hussars—who thus gladly acquired another Prince Albert. The official designation of the regiment was now changed from *11th (Prince Albert's Own) Hussars* to *11th Hussars (Prince Albert's Own)*. Prince Albert was not to see his regiment again for some time, as between October 1921 and January 1926 it was in India, but a deputation of all ranks attended his wedding at Westminster Abbey in 1923 to Lady Elizabeth Bowes-Lyon.

Between 1926–1934 the regiment was back in England, stationed successively at Shorncliffe, Aldershot and Tidworth. Since 1924 the designation 'Private Soldier' had been styled 'Trooper', a change long overdue in the sense that the Troop had always been the vital formation in the cavalry. But the day of the horse in warfare was running out. By far the most important event that took place during these nine years was the conversion of the regiment—at the same time as that of the 12th Royal Lancers—from horses to armoured cars, and the subsequent learning of how to use these strange new steeds. Although in after years the two regiments looked back with relief, and indeed with pride, on their selection as the first cavalry regiments to be mechanised, it actually came about because they were the two most junior cavalry regiments that had not been amalgamated since 1919. At the time that the news was broken, however (March 1928), it came as a shattering, almost unbelievable blow. For the 11th had been a regiment of horsemen for 213 years; they had trained with, lived with, and fought with horses. If the good infantry officer saw that his men were fed and cared for first, the good cavalry officer's order was first horses, second men, third self.

Major-General Pitman, whom the Army Council had most wisely asked to break the news to his regiment, took exactly the right line in doing so. It was indeed a bitter blow, he told the officers, but if they thought about it further they might see a great opportunity before them. For the role of reconnaissance had always been a perquisite of the cavalry, whose value had always lain in its mobility. It was sad but true that modern developments meant that the horse must give way to the machine; yet here was a great chance for the regiment to show itself, whether horsed or mechanised, still the most efficient in the Army. Lieut-Colonel F. H. Sutton, M.C., Commanding Officer since 1924, who was to have the first of many headaches as mechanisation became a reality, had already told his officers: 'Either we all go or we all stay.' There was not much doubt after that.

On April 10, 1928, was held the last mounted parade of the regiment. A few days later nearly all the horses reached Melton Mowbray whence the Inspector of Remounts wrote to Colonel Sutton congratulating him and the regiment upon their condition: 'I have been Inspector of Remounts for seven years, and I have never seen horses returned anything like so good as the Eleventh Hussars' were.'

Armoured cars in 1928 were by no means a novelty, for they had been used on occasion during the war with success, and thereafter ten armoured car companies had been retained with the Royal Tank Corps. The Royal Air Force also had long been interested in their use for ground defence and observation. But compared with even the Humbers and Daimlers of thirteen to fourteen years later they were primitive in the extreme. The trouble was in 1928 and 1929 that they also appeared to be extremely scarce, for there was great delay in completely equipping the regiment with its cars long after all ranks had completed their driving and maintenance training and innumerable courses in machine-guns, revolver, wireless, tactics, gunnery and what-have-you. For much of the benefit of these the regiment stood greatly in the debt of the 12th Armoured Car Company and the 3rd Tank Battalion.

Lieut-Colonel A. L. I. Friend, M.C., had succeeded 'Squeak' Sutton soon after mechanisation, and writing in August 1929 he noted that the first new Lanchester had arrived on June 25, 1929, followed three weeks later by another, old one.

The Rolls-Royces, which had started arriving in January 1929, had been designed originally in 1917, and with variations and improvements in design (the '1920–1924' pattern) were, unbelievably, to constitute the main striking strength of the regiment until 1941. They were lightly armoured, carried a crew of three, one of whom sat in the turret and operated a single Vickers machine-gun on a ball mounting. In the second World War new two-man turrets were fitted and the armament was also increased to a ·55 Boyes rifle and a Bren gun. It was not until 1935 that wirelesses were fitted on the scale of one per Troop of three cars. As always, the record of performance and reliability of the Rolls was splendid (some cars that the regiment used against Graziani's armies had chassis used in the Great War), but they were not good in soft sand and drank water —both desert liabilities. And, naturally, in 1940 they were obsolete before they began as regards both armour and armament. But they are still recalled with great affection.

This is to forestall events. In 1929 Colonel Friend's establishment also included sixteen two-seater Austin 7s (forerunners of the Jeep?), thirty Triumph motor-cycles, and some Crossley and Leyland six-wheelers. Altogether the regiment had 117 vehicles of one kind or another.

It can be seen that a good deal of trial and error was of necessity going on in the regiment, as in the Tank Corps, due presumably to an equal if not greater amount of backing and filling at the War Office or in the Army Council. One question that had not yet been resolved in August 1929 was the form of headgear to be worn by the car crews. Colonel Friend had suggested a beret of 11th Hussars design, but the War Office, which originally had thought a beret of some sort the obvious answer, suddenly came out with a new form of cap of its own design. Eventually, as might perhaps have been guessed by an interested outsider, the 11th were allowed to wear their own beret of brown with a small crimson band.

Besides the interesting new problems set by new vehicles, new wireless and R/T training, and new weapons, problems which applied to everybody—for early Colonel Friend had wisely laid it down that as far as possible every man should be capable of doing every other man's job—there was one department which underwent more radical changes than any one other. This was the Quartermaster's, still in the Stone Age, as it were, of mechanisation. Even so when the Q.M. had to provide 1,800 gallons of petrol (in those days a huge amount) for an exercise in 1929 it was quite an achievement that it all duly turned up, thus enabling the armoured cars to cover 200 miles each during three days. That mileage in itself without many breakdowns was reckoned a considerable achievement. The days of a permanent Light Aid Detachment and specialised Squadron fitters within the regiment were still some way off.

It was one thing for officers and men to become confidently competent with the armoured cars; it was quite another to learn—and teach—the best tactical way of using them and the right formations. As no cavalry armoured car regiment had ever existed before, this problem had to be thought out from scratch. No small debt is owed to Colonel Friend, for it was he who largely put forward the proposals which solved the various problems and which were adopted by the War Office. The use of armoured cars as adopted in the Royal Tank Corps did not sit happily within the framework of a cavalry regiment (numbers of and in Troops, for example, differed, and the Tank Corps used Companies of cars); Colonel Friend proposed that cavalry regiments with armoured cars should be organised into squadrons (usually three plus headquarter squadron) with three cars in each Troop (usually three, later four per squadron). This and other proposals were accepted. In effect, therefore, the 11th Hussars wrote the new *Armoured Car Manual for*

Tactical Handling and Drill for the 1930s, for the 12th Lancers were in Egypt and thus comparatively immune to immediate blasts or zephyrs from the War Office.

Between 1930 and 1934 the regiment was at Tidworth, and at the end of those four years after many false starts, false trails, aches and pains a complete cavalry armoured car regiment had emerged, the first of its kind in England. Its development had been watched with close interest from on high, and the inspection reports showed that a first-class and potentially very valuable unit had been developed. Colonel Friend had been succeeded by Lieut-Colonel D. McM. Kavanagh, the machine-gun expert of the past war, in 1932. Major-General Arthur Friend's great administrative gifts were to prove equally valuable in the raising, training, and exploitation of the Pioneer Corps, which he commanded in the 1939–1945 War. Certainly the regiment had reason to be grateful to him in its difficult transition period.

At Tidworth the 11th were to see much of H.R.H. the Duke of Gloucester, whose own regiment, the 10th Hussars, had gone abroad. While on Staff duties in 1931 and 1932 he lived in the Officers' Mess, and for over twenty months in 1932 and 1933 he was actually attached to the regiment and commanded 'C' Squadron. In later years one veteran trooper of those days used to claim that he had been punished by the Duke (for having a dirty rifle) in the following terms, when he had been marched in for squadron orders.

'Seven days to barracks. That'll teach you to bugger about with my old man's army!'

But there were to be many happy associations with the Duke of Gloucester, who has always looked upon the 11th as his second regiment.

The many lessons learned at Tidworth and on Salisbury Plain during these years were valuable, but some of them were unrealistic, and other lessons were not to be had under English conditions. Lieut-Colonel A. P. Wavell (later Field Marshal Earl Wavell), although he had been G.S.O.1 of an infantry division for three years between 1926–1928, had been pondering the use of armour more than perhaps appeared. For he suggested to a Staff College conference in 1929 that the respective value of an armoured force's assets of mobility, fire-power and armour might be put at the ratio of 3:2:1. Major-General R. J. Collins, in his biography of *Lord Wavell* (1883–1941) adds:

> 'He put his finger on the main weakness of such forces, of which those handling them were only too painfully aware already, namely their sensitiveness to ground and thus the desirability of a small force able to fight on foot being incorporated in such armoured forces so as to enable them to secure a crossing over a river or a passage through a defile.'

Collins himself had been pleading for an infantry battalion to be included in even a small armoured force, but in those days his plea 'fell on deaf, not to say shocked, ears'.

Nobody envisaged at this time that the first great British use of armour in war would be in the Western Desert of North Africa. But Wavell and others, much as they had learned from exercises on Salisbury Plain, then as now the only area in Britain where tracked vehicles could be manoeuvred on a large scale without raising the rooftops of rural and public protest, realised that the area was still far too circumscribed for true practical training with large armoured forces. False lessons in fact could easily be learned, especially regarding armoured cars. Collins writes:

> 'At the time he stressed that the main role of such fast-moving vehicles, largely tied to roads in England, was strategical reconnaissance, and that they would in war seldom come into the same picture as the light tanks with which they were then grouped and usually oper-

ated. The organisation of an armoured division in the Second World War, with its Reconnaissance Regiment, showed how sound his thinking was and how right were his deductions. Many years later, under his command, the 11th Hussars were to show just how such units should be employed and handled.' (Wavell, of course, had also stressed the vital importance of developing wireless communications, since the wider the front on which armoured forces could move, in his view, the more formidable they would be.)

In 1934 the peace-time establishment of the regiment at home was provisionally laid down as: R.H.Q. and three Squadrons, each of headquarters and three Troops; twenty-one officers, thirty-four N.C.O.s and sergeants, 380 rank and file—total all ranks 435. Armoured cars thirty-eight, motor-cycles thirty, lorries and other vehicles thirty-one.

When the regiment arrived at Helmieh, Egypt, in November 1934 its 'Colonial Establishment' was very much the same, but it was in fact nearly 100 under strength in other ranks. This deficit was largely made good by a draft of seventy 12th Lancers, the outgoing regiment.

Until 1922, the year of the first cavalry amalgamations, the 11th and the 13th Hussars had been affiliated, but then the 13th and 18th Hussars joined together. Since mechanisation the regiment's closest affiliation had obviously been with the 12th Lancers, and for a decade each regiment supplied the other with small drafts from time to time, or there would be cross-postings. In 1929 the regiment became affiliated with the 1st Canadian Hussars, which had grown out of a volunteer troop of cavalry formed in London, Ontario, in 1856. In 1872 that became the First Regiment of Cavalry, and in 1892 the 1st Hussars. This fine Canadian regiment was to serve thereafter with distinction in the Boer War and both World Wars.

When individual cavalry regiments were linked with individual yeomanry regiments, in a sort of 'parent-brother' relationship, the 11th formed the close association with the Royal Gloucestershire Hussars—alongside whom it was to fight sometimes in Africa—which turned out to be so happy and profitable for both. By regularly providing an Adjutant, Quartermaster, sometimes a second-in-command, and always various experienced and valued senior N.C.O.s, the regular half of the partnership has always been able to help and direct the enthusiasm of the Territorial half.

In the nine years of home service between 1926–1934 the 11th Hussars continued to polish its reputation for weapon training. Although only sixty rifles were left on mechanisation, this did not prevent the regiment from winning the Cambridge Shield for the best cavalry regiment at shooting for six successive years, a record that has never been equalled. At this period Peter Payne-Gallwey was at the height of a successful career as a steeplechase jockey. Of many successes his winning of the Grand Military Gold Cup at Sandown in 1933 on his horse 'Backsight' stands out. Another future commanding officer, Captain W. I. Leetham, at this time away from the regiment in India, rode many winners there between 1926 and 1932, and won the Indian Grand National.

It was a very changed regiment, both in men and mounts, that sailed from Southampton to Alexandria on November 15, 1934. It was to be over nine momentous years before it saw England again.

In Egypt the regiment had thirty-four Rolls-Royces and five Crossley armoured cars—the Lanchesters, never really satisfactory, having been superseded. The Crossleys, which had wireless, were used as headquarters and command cars; they in turn were replaced in 1938 by Morris armoured cars. The latter had been designed to cope with soft sand, which they did well enough; they were very lightly armoured and had a two-man turret with the same armament as the later Rolls-Royces—which they were supposed to replace. But in 1938–39 only thirty Morrises reached the regiment, at a time when squadrons were being reorganised on a five troop basis, so most of the old Rolls were kept and the Morrises were taken as replacements for the Crossleys. Thus when the regiment went to war in June 1940 each troop leader had one Morris, with wireless, and two Rolls. All the lorries in the regiment by then were also Morrises. But even in April 1937 Brigadier Friend, Commander, Cairo Cavalry Brigade, wrote of his old regiment in the annual inspection report:

> 'Armoured cars old and obsolete but well maintained. I hope to see all transport of one type by next year and the six-wheelers abolished for this country.'

He was referring to the Rolls as obsolete; but they and the Morrises were to give good service until the spring of 1941, although by then there were only a handful of desert-worthy vehicles left.

It is worth remembering that even in late 1935 the attachment of an R.A.O.C. Light Aid Detachment was a novelty to the regiment, although by now it had become accustomed to and very grateful for a permanent section of the Royal Corps of Signals integrated within the regiment. These essentials were to be magnified and perfected to an almost unrecognisable shape by the end of the war, and thus tended to be taken for granted.

On the regimental side the first necessity from 1935 onwards was to know and make the very best of all one's equipment, however inadequate it might often seem and however much more difficult the task became with changes and deficiencies. The second, not less important, requirement was to get to know how to make the best use of it tactically and in fighting; and the third, certainly no less important than either of the other two, was to get to know the desert, much of which was unmapped or even totally unknown. It can be seen that thorough training based on the assumption that all three requirements interlocked was essential, and the 11th were fortunate to have several years in which to perfect their abilities before war broke out, and equally fortunate in the Colonels who trained them. A fourth requirement, or less a requirement than a basic fact of life, was a really efficient Quartermastering department. That the regiment had always had, but totally new problems of staggering difficulty were raised by the desert, which if it was to turn out the tactician's paradise was certainly also to be the Quartermaster's nightmare.

General Collins, in his book on Wavell, writing of the years between 1936 and 1940, records that

> 'desert reconnaissances had been carried out by all units westwards right up to the "Wire" and southwards as far as the Siwa Oasis. As a result a confidential coloured map had been compiled by means of which the surface of the desert, and the type of vehicle which could traverse it, could be deduced at a glance. The 11th Hussars, the Reconnaissance Regiment of the 7th Armoured Division, were the pioneers in this work and set a very high standard of all-round efficiency.' (The 'Wire', an entanglement thirty feet broad and six feet high, ran from the Mediterranean due south for 150 miles marking the frontier between Egypt and Cyrenaica, part of Mussolini's empire.)

All 11th Hussars who served early in the desert agree that the regiment—and not only the regiment—owe an outstanding debt to Colonels John Galbraith and, above all, John Combe. It was due to their foresight and tireless efforts that the battle was half won already against the Italians when war came to Africa in June 1940.

In fact the 1935 Abyssinian crisis, during which Mussolini moved Libyan troops to the Wire, made Britain sit up. It also revealed lamentable deficiencies in her warlike stores and capabilities, not least in the Middle East. A so-called 'Mobile Force' was hurriedly botched up in Cairo to defend the Western Desert if need be—and it

The 11th Hussars in the Coronation Procession, May 12, 1937.

looked very likely that need would—and trundled out amid rain, mud and then sandstorms to the Mersa Matruh area. So many tank tracks got clogged up and so much transport bogged down that this rather rusty spearhead soon became known, unkindly, as 'the Immobile Farce'.

But, as Brigadier Clarke comments in *The Eleventh at War*,

> 'The crisis did at least bring in its wake one blessing, for which we were later to be thankful. As nothing else had ever done before it cast a glaring light upon the flaws in our defences, showing up the gaping holes where they had been eaten away by years of parsimony and the complacent delusions of past governments. For the Army it produced a sudden speeding in the leisurely process of mechanization. . . .'

War, however, did not come at this stage. But for many of the British troops in Egypt, including the 11th, action soon did—in Palestine in 1936. In retrospect this seems to have been fortunate in one respect, that it gave the regiment and its friends a chance to see how they behaved and if their training really worked when things became rough. For the great majority of the troops had never been in action. The Palestine troubles were to be an unwanted but useful way of easing them towards professionalism at war, if this is not too cynical.

The Arab Rebellion that burst out in April 1936 just after Easter was the result not just of the rapid rise in Jewish immigration, as previous revolts directed against the Jews had been, but also of dissatisfaction with the British, whose prestige slipped heavily during the Abyssinian crisis. When attempts to smuggle arms and ammunition into Tel Aviv came to light, a full-scale rebellion was bound to come, and its actual occasion was a series of racial murders and reprisals. As usual, the first sparks were struck on the borders of Jaffa and Tel Aviv, and an Arab general strike followed immediately.

The R.A.F. was responsible for military security, and all British military forces came under the command of Air Vice-Marshal R. E. C. Peirse; besides his own two squadrons of bombers he had only one R.A.F. armoured car company and 900 men of the Transjordan Frontier Force. 'British Troops in Palestine' was a grandiose title for two infantry battalions—the 1st Loyals and 2nd Queens Own Cameron Highlanders—and No. 14 Company, R.A.S.C. Para-military forces available were about 2,000 Palestine Police, over half of whom were Arabs.

The 11th did not arrive in Palestine until early July 1936, when the last stage of the rebellion, regular guerilla warfare, had been reached. They were allocated to the northern of

Windmill Hill Camp, Salisbury Plain. 'C' Squadron on annual manoeuvres in 1930 equipped with Lanchester and Rolls Royce armoured cars.

two brigade sectors, and R.H.Q. was established at first in the old Turkish fort of Nablus. At that time there was no direct coastal road linking Tel Aviv with Haifa, so that all north-south traffic had to run the gauntlet of the central hill massif, completely controlled by Arab guerilla bands. Over 100 miles of this route could be negotiated safely only with strong military escort. Even so, such convoys were frequently ambushed. The British object was always to bring the rebel bands to action and pin them down, preferably by daylight. It was the first experience of guerilla warfare for the 11th and they found it fascinating. Its greatest value to the regiment was that it not only enabled but compelled squadrons and Troops to operate on their own. Thus all commanders from squadron leaders down had unusual responsibility and gained experience and confidence; they learned to be separated by many miles from their friends and headquarters, and they learned to think clearly and act decisively while bullets were flying around or mines exploding. Most valuable of all perhaps was this foretaste of war for the individual Troop leaders—some of whom were sergeants—and their car crews.

The regiment provided escorts for regular day-time convoys, for mobile striking forces waiting in reserve for the wireless code signal, usually from the R.A.F. mobile signals unit, and for roving artillery patrols by night. These latter used naval 2-pounder pom-poms or 3-pounder quick-firers removed from the decks of Royal Navy warships and installed in lorries hired from a Jewish firm!

R.A.F. armoured cars looked after the southern zone, but in the northern there were heavy demands on the forty cars of the regiment. 'A' Squadron from Nablus covered the main road from the bottom of the northern zone to Deir Sharaf; 'B', working from the inflammatory town of Jenin, took over from Deir Sharaf to Affula, with a detached Troop at Tulkarm; and 'C' was responsible for the northernmost roads in the Galilean hills.

R.H.Q. soon moved to Haifa, and 'A' Squadron at Nablus, which appeared to have a peculiar attraction for snipers, soon found the most ever-present danger to be the land mine. These were usually old Turkish shells dating from the Great War, filled with scrap iron and gelignite, with a hole drilled in the base for the detonator which had a nail inserted in it; to this was attached a sort of mousetrap to which string or wire was tied. The whole rather Heath-Robinsonish but very effective device was placed in a hole in the road or ground with the string or wire stretched only a few inches above the earth. One of these mines killed a Seaforth Highlander officer in a Morris truck, and wounded three of his men, near Nablus, after an 11th patrol had been sent back. When it rejoined, Lieutenant Wainman's car set off another mine but luckily no one was hurt.

On September 3, 1936, a big engagement took place at Kilo 88 on the Nablus–Tulkarm road, near 'Windy Corner' as it had become known. One R.A.F. plane was shot down by Arabs and its occupants killed, another, whose pilot was wounded, had to make a forced landing. The infantry suffered three killed and eight wounded, but the Arabs far worse casualties. During the fighting, which went on all day, the ambulance from Nablus had been very busy, and on its return trip at night it was fired on; Lance-Sergeant Harry Petch, commanding two cars, was sent out to escort it to safety. He saw the string of a land mine in the beam of his headlights, and got out of his car to investigate. As soon as he did so he was fired upon, but despite the fire being kept up he calmly disarmed the mine and put it in his car. Then he safely brought home the ambulance. Sergeant Petch, who had joined the regiment as a Trooper in 1930, was awarded the D.C.M.; he was to end the war as a Major and Squadron Leader, with an M.C.

'B' Squadron had an exciting day on August 21 when with a platoon of Loyals they went through the Musmus Pass to Ara to hunt up survivors from a skirmish the day before. On their way back they were ambushed in the pass by over sixty Arabs, well concealed in thick scrub and woods. R.A.F. aircraft were called in and caused twenty-six casualties with bombs and machine-guns.

At the end of September the 1st Division arrived from England, followed by the 5th, as the British Government realised it could not afford the drain of troops from Egypt as Arab resistance got stronger. The whole garrison was put under command of Lieut-General J. G. Dill. On October 12, as a result of these reinforcements, the Arab strike, which had lasted six months, was called off, and convoy escorts became unnecessary. The first time that a convoy went unescorted through the somewhat sinister town of Jenin, a deputation of Arabs came to 'B' Squadron's camp to complain that passing Jews had shouted insults at the Muslim religion. The following conversation took place between the Arab leader and Lieutenant Robin Stuart-French, who received the deputation with his usual grace.

'I want to see the General.'

'There isn't any General here. I'm the senior officer.'

'Well, if there isn't any General, I want to see the Sergeant-Major!'

At the end of October 1936 the 11th Hussars took the long road southwards across the Sinai desert to Helmieh. For the next twenty months they stayed in Egypt, but there were frequent training exercises and manoeuvres, some of them on a large scale, so that all the time co-operation with other arms, practice in the functions of an armoured reconnaissance regiment, and knowledge of the Western Desert, where most of the collective training was carried out, were increased. In particular the regiment built up a closer and warm relationship with the R.A.F., already started in Palestine, especially with No. 208 Army Co-operation Squadron (Audaxes, later Lysanders), soon to be regarded as almost a permanent part of the 7th Armoured Division. It became axiomatic for officers and sergeant-pilots of this fine squadron to travel from time to time on the armoured car reconnaissances, while in turn officers of the regiment were given frequent air trips to see what the desert looked like from above, both by day and by night.

But in July 1938 the 11th were again back in Palestine, where a new revolt had broken out, and stayed for three months, individual Troops being even more widely dispersed than before. Colonel Galbraith was expecting to stay in Palestine indefinitely, since the state of rebellion got worse rather than better, when because of the Munich crisis in September 1938 he was directed to bring back the regiment post haste to Cairo. Eventually 'B' Squadron was left behind. The Munich crisis resolved itself—temporarily —and amid general jubilation the 'flap' in London died

Helmieh Camp, Cairo 1937. Col John Galbraith leads the 11th Hussars on parade at the Annual Inspection.

down. In Cairo precautions were taken, however, and when Colonel Galbraith sent 'C' Squadron and an advanced headquarters back to Palestine in October he was ordered to leave a large cadre behind in Cairo for training recruits. These latter tours of duty in Palestine may have been less exciting than the first, but they were arduous as well as usually tedious, despite the usual incidence of ambushes and skirmishes. In April 1939 Mussolini invaded Albania; from Cairo the British garrison sped out for a third time towards the desert frontier, and for the third—and last—time the 11th were recalled to Egypt from Palestine.

Major-General B. L. Montgomery, of whom more was to be heard, was commanding 8th Division in the northern zone of Palestine at that time, and the 11th provided a standing escort for him. True to form—although his form was an unknown quantity to the 11th in those days—'Monty' gave every member of the car crews a signed photograph of himself as a souvenir.

In July 1939 Colonel Galbraith handed over command of the regiment which he had brought to a high state of efficiency to Lieut-Colonel John Combe, who had already contributed much to its progress and in the year left before it saw fighting in the second World War was to contribute even more. In 1939 Major-General Pitman, who had been an active and helpful Colonel of the regiment since 1926, handed over to Brig.-General Sir Archibald Home, C.B., C.M.G., D.S.O. The Colonel-in-Chief was still of course the same, except that he had been crowned King George VI in 1937.

On August Bank Holiday, 1939, the 11th left Helmieh once again for 'special training' at Mersa Matruh. This fooled nobody, as war was thought to be very near. It was declared on September 3, 1939, while the regiment was in camp beside the peaceful blue lagoon and white sandy beaches they knew so well from previous years.

Chapter 12

The Western Desert: The First Offensive

'Their exploits will go down to history, but I doubt whether history will ever be able to do justice to the cheerful spirit of quiet efficiency that was the hall-mark of the Division in good times or in bad. I never visited the 7th Armoured Division without coming away heartened and impressed by this spirit, and I know that many others of varying ranks, including some of different nationalities, have been similarly impressed.'
(Letter of General Sir A. P. Wavell, C-in-C Middle East, to Major-General M. O'Moore Creagh, February 25, 1941.)

THE 11th Hussars fought in North Africa from June 11, 1940, the day when Italy declared war, until May 9, 1943, when all Axis resistance there ended. Except for a few brief spells in Cairo and elsewhere for rest and refitting, and one short (in time) trip to Iraq and Persia, they were continually in action. Very often, if not usually, squadrons were widely divided or operating on their own for periods of time dictated by the course of the battle; yet wise, authoritative and perceptive command from Regimental Headquarters was not only reassuring but vital, since the tides and eddies of desert warfare could behave surprisingly. Individual Troops too were often alone for days at a time. Leaders from the Colonel down often had to make difficult and dangerous snap decisions. One cannot realistically talk about a 'front line' in an area as vast as the Western Desert or Tripolitania, and in desert warfare there was normally no such thing as a clearly defined front.

There were, however, groupings dictated by geography and tactics which became familiar, and in the sense that the 11th usually knew where the major and minor strengths of the enemy were (although not always), and certainly knew the enemy's dispositions sooner than anyone else, there can be said to have been 'enemy lines'. Right up to El Alamein in 1942 the 11th were concerned very much with the front-line soldiering all the time—not that they did not have plenty after Alamein, but by that time the desert was getting crowded! Furthermore, for the first two years 11th patrols were constantly operating behind the enemy lines in the sense mentioned above, and very often many miles behind them.

Until a definitive history of the regiment is written, therefore, it is quite impossible to do justice to all the work that was done by it in Africa. There is simply too much material. What follows, then, are several selected examples of what happened, at different periods, including an account of one great but neglected victory and one grand but hardly neglected finale. Just as those wishing to know more details about the regiment's part in the Great War can consult the late Earl of Scarbrough's history, so those interested in its desert story should read the very full account given in Brigadier Clarke's *The Eleventh at War*, and also the semi-official and unofficial histories of the 7th Armoured Division. If little attempt here is made to present 'the big picture', that is because it is already so familiar to many, so readily available in countless books to all.

It is roughly 550 miles from the white dunes of El Alamein to Benghazi, as the crow flies, and from Benghazi the Mediterranean coastline dips sharply southwards in a scoop to Agedabia and El Agheila before continuing west-north-west through Tripolitania to Tripoli and Tunisia. From Gabes on to Tunis it is almost due north. And Tunis is something like 2,000 miles from Cairo by the coastal road.

But most of the fighting took place between Sidi Barrani and Agheila, and a great deal of it in Cyrenaica, ten, twenty, thirty, fifty, sometimes even 100 or 120 miles from the sea in an area bounded by, roughly, Agedabia, Benghazi, Derna, Gazala, Tobruk, Bardia and Sollum at the coastal end and by the open desert—stretching for mile after mile down to the great Sand Sea in the south. Only Giarabub and the Siwa Oasis are names near the Sand Sea that mean anything. The Sand Sea itself set a limit to operations, being impassable, like the Quattara Depression in Egypt, and in the north between Benghazi and Derna lay the huge expanse

Imperial War Museum

The 11th Hussars in N. Africa. The indifferent sun induced a very curious array of headgear.

of the Jebel Akhdar (Green Mountains), full, it is true of scrubby woodland and water, but off the tracks unsuitable for mechanical manoeuvre.

In all this area there was in fact only one good road, the coast road. Clearly problems of supply alone dictated that most of the fighting would not be more than 100 miles from the coast, and probably less, since the various ports were vital. The Sand Sea is usually about 175–200 miles from the coast. The most popular or populous battle arena was the sandy coastal plain that stretches for the most part to a width of thirty miles, bounded by the Escarpment, which, however, narrows it down to a few hundred yards at Sollum where the Escarpment comes down to the sea.

At the height of the Munich crisis Major-General P. C. S. Hobart had been flown from England to Egypt with orders to 'create an armoured division'. No man could better have been suited to this task. The Mobile Division (Egypt) which he formed from the old Mobile Force in 1938–39 was an altogether different proposition. Its equipment had been somewhat improved, though still primitive by the standards of even two years later, but it was far better trained, and its administration was completely reshaped. Indeed the system that General Hobart and his D.A.A. and Q.M.G. laid down became the basis for all mobile formations in the desert later. It was sadly ironic that General Hobart, an experienced, brilliant, but sometimes difficult officer, never commanded the 7th Armoured Division, his own creation, in battle. His enthusiasm had never flagged.

When war broke out the Mobile Division consisted of the Light Armoured Brigade (7th, 8th and 11th Hussars), the Heavy Armoured Brigade (1st and 6th R.T.R.) and the Pivot Group (3rd R.H.A., 'F' Battery, 4th R.H.A. and 1st K.R.R.C.).

It soon became evident that Italy was going to sit on the fence for the time being, so while the R.A.F. watched the Libyan frontier more training continued, and some new equipment arrived. Hobart's departure in the winter of 1939 came as a shock to the division, but after Major-General M. O'Moore Creagh had taken command of it early in 1940 it soon reassumed its identity, although now known as 7th Armoured Division.

It was part of Lieut-General R. N. O'Connor's Western Desert Force, which could call for air support only on Air Commodore Collishaw's pitifully small group of several Flights of Gladiator fighters and Blenheim bombers, and 208 A-C Squadron—nucleus, all the same, of the future Desert Air Force.

By the time Italy declared war the Light Armoured Brigade had become 7th Armoured Brigade, the Heavy, 4th Armoured Brigade. 1st and 6th R.T.R. had swopped places. Only one squadron of the 8th Hussars in 7th Armoured Brigade so far was properly equipped with tanks. The Pivot Group became the Support Group and had 2nd Rifle Brigade added to it; 3rd R.H.A. were converted entirely into an anti-tank regiment with 37 mm. guns; and for all major purposes 4th R.H.A. with 25-pounders became part of the division. But there were no heavy tanks; only light and a few old medium ones, until from October 1939 onwards new cruisers started to arrive, but in derisory numbers.

Shortly before war with Italy the 11th had been reinforced by fifty-five Rhodesians who were to do splendid service. It also had with it when it crossed the Wire on the night of June 11, 1940, four officers of the R.T.R., who became 11th Hussars in every sense.

Colonel Combe's orders on June 11 were to penetrate and where possible destroy the frontier wire, to 'dominate' the area between Forts Capuzzo and Maddalena, to delay any Italian advance, and to harass their communications back towards Bardia and Tobruk. The first engagement of the desert war was when Major Geoffrey Miller and his two 'B' Squadron headquarters cars ambushed and captured four Italian lorries about 2 a.m. on June 12; they were going from Fort Capuzzo to Sidi Omar. This little affair yielded two officers and sixty-eight other ranks besides three Italians killed.

Many of the Italian frontier posts seemingly had not known when war was declared; their occupants rapidly did, some being caught, so to speak, with their trousers down. Soon they were all in British hands, mostly captured by the regiment. Fort Capuzzo was the exception. On June 14 it was taken with over 200 prisoners by the 7th Hussars and a company of K.R.R.C. after 'a dramatic scene in which those Cruisers, which were lucky enough to have guns, bombarded the walls of the fortress with 2-pounders'.* The fort was demolished, but throughout the summer fighting took place round it and Capuzzo changed hands several times. The same day 'A' Squadron captured Fort Maddalena after some erratic bombing by a flight of Blenheims; surprisingly this fort, although an elaborate affair easy to defend, contained by then only five Italians and thirteen Libyan soldiers. But on June 13 when T.S-M. E. S. N. Clarke and his Troop had driven almost up to the front door —within 500 yards of the main gates—they had been met by concentrated fire from what turned out to be a dozen machine-guns; then they had been chased back across the wire by six Caproni bombers and nine Fiat fighters. At this time the Regia Aeronautica had complete air superiority, and time and again were to make life very uncomfortable for the regiment. The wireless station at Maddalena was destroyed. On June 13 'A' Squadron had also taken posession of Sidi Omar, which the Italians had evacuated in haste, although it had been defended two days before; Lieutenant J. A. Friend set fire to its two main forts.

General O'Connor now knew that the Italians were

* *A Short History of the Seventh Armoured Division* (October 1938 to May 1943), by Lieut.-Colonel R. P. M. Carver, D.S.O., M.C.

showing no inclination to advance eastwards along the coast road, and felt that the initiative was in his hands. Fort Capuzzo had been the strongest Italian position on the whole front and the likeliest base for any attempted invasion of Egypt. After its initial capture, however, a regular hornet's nest was found at Sidi Azeiz, north-west of Capuzzo, which protected the road to Bardia. This was a strongly fortified and defended camp, protected by a minefield, as the 7th Hussars found to their cost when 'Combe Force', composed of one squadron of their tanks, two 11th squadrons, and one R.H.A. Battery, attacked it on June 14. Three tanks were blown up immediately, and the rest immobilised for some hours. One lesson was brought home, so basic that it seems absurd now; the 7th had No. 11 wireless sets, the 11th No. 9 or No. 16 sets, so the two could not communicate with each other. A reflection indeed on British unpreparedness for war.

The Italians had two craftily concealed gun batteries some way behind the main ridge protecting Sidi Azeiz, and the only way to capture the objective was to knock these out first. But when the sun was up the desert danced with mirages and the glare made it almost impossible to pick out these targets. 'C' and 'B' Squadrons drove their armoured cars right round to the rear of the defences drawing enemy fire so that the R.H.A. could spot the flashes, but all afternoon the Italian guns dominated the approaches, even to the extent that a large Savoia bomber felt it safe to land inside the perimeter of Sidi Azeiz.

Colonel Combe decided therefore to leave 'B' Squadron to keep guard, draw off the guns and tanks to the south, and bypass the enemy, since he could only thus fulfil his other task of reconnoitring Bardia. He sent forward 'C' Squadron towards Bardia. Major Payne-Gallwey's first problem was to find a way down the face of the Escarpment to the metalled road below. A passable track was soon discovered, and the squadron debouched on to the coastal road—empty as far as the eye could see. But Bardia turned out to be strongly held with all approaches covered by infantry in trenches. Towards last light the squadron made for 'home' —in this case one of the gaps in the Wire. On their way, however, they were astonished to see smoke and flames rising from the direction of Sidi Azeiz; and getting closer a white flag was observed on the post's observation tower. Major Payne-Gallwey naturally headed for the garrison, for in the distance an Italian convoy was making for Bardia; but when his squadron was 500 yards away from the Italian trenches a hail of shell-bursts fell among them, fortunately without hitting any car, and the white flag was lowered. 'C' Squadron got away safely, but were pursued to the Wire by nine light tanks. Had they but known, these were probably not very menacing, for earlier Second-Lieutenant C. A. Halliday of 'B' Squadron had suddenly run into six which attacked his Troop, and had knocked one out with a Boyes rifle. Obviously these Fiat-Ansoldas were very thinly armoured. Soon the 11th did not fear them.

On June 16 'C' were back again on the Tobruk–Bardia road, having leaguered the night before on the edge of the Escarpment, some twenty miles behind the front Italian positions. Here they arranged a most effective ambush system and had a splendid day, for none of the Italians travelling from Bardia to Tobruk or vice versa had any inkling of the presence of British troops. Without any casualties they captured forty vehicles, mostly lorries, nearly 100 prisoners, and killed twenty Italian soldiers. One of their 'bags' was rather more important than the rest— three staff cars containing General Lastrucci, chief engineer of the Italian 10th Army, and his staff, not to mention, to the great delight of the 11th, two lady friends. The General had with him some excellent marked maps of Bardia's defences. The squadron was attacked once by fighters which hit one car mildly, but at dusk it made for the open desert, where things always felt safer, and by dawn on June 17 was back across the frontier with all its booty.

Valuable and stimulating as such successes were—and they were to become very typical—a much more important victory had been achieved elsewhere the same day, June 16. This was the first tank skirmish of the war, near Ghirba. Here a mixed Italian force of twelve light tanks and thirty lorries carrying infantry, all bound north for Fort Capuzzo, was sighted by Vivian Gape near the frontier. It was early morning, and soon afterwards George Dier reported forty lorries and 300 infantry moving west not very far away, escorted by seventeen light tanks. The two columns looked like converging, so Major Miller gave his two Troops orders to withdraw and asked Colonel Combe to get some anti-tank guns up from 4th Armoured Brigade.

Some of 'B' Squadron, however, had already moved into the attack with two armoured cars and raked the lorries with machine-gun fire, bringing the first column to a halt. Gape then engaged its tanks in a running fight, knocking out two of them. One of the Rolls was slowed down by a hit which punctured a wheel, and the tanks were closing in as the Troop commander hung back to protect it when help arrived in the shape of Dier's Troop which had driven towards the firing. Another tank was destroyed and the rest turned back to rejoin their infantry, who by this time had brought a field gun into action. The 11th took cover behind a ridge and reported back to Colonel Combe on Gape's wireless. Both Troop commanders were in a nasty position with only five cars between them. Combe happened to be at 4th Armoured Brigade headquarters where Brigadier Caunter placed under his command a mixed tank squadron of the 7th Hussars under Lieutenant D. Seymour-Evans and an anti-tank Troop of 3rd R.H.A. These met him at the Wire and after ordering two Troops from 'A' and one from 'C' Squadron of his own regiment to join the fight Colonel Combe held a quick conference at the 11th's position.

For no logical reason the Italians had formed their lorries and infantry into a solid square in the open, with their tanks on the sides rather like destroyers ranging up and down a convoy. The 7th had only five tanks, and had to use a Morris of the 11th to communicate. But they went boldly in line ahead at the Italian square (their starting point was three miles distant) whereupon seven Italian tanks came out to meet them. The British tanks shot them to pieces in a few minutes, and then circled the square pouring heavy fire into the infantry.

The Italians, however, had a trick up their sleeve in the shape of a field gun at each corner of the square, and these, bravely manned by Eritrean native gunners, fought a close-range duel with the British tanks. Finally all the gunners

Imperial War Museum

N. Africa. There is little or no shade in the desert; the 11th brought their own.

were killed or wounded at their guns and the Italian infantry began to panic as several ammunition trucks blew up. They tried to escape in lorries but were pounced upon by the armoured cars. The Italian light tank crews fought as bravely as their gunners but they were no match for the 7th, who knocked out several more: the 11th accounted for another five. By early afternoon the fight was over. Without a single casualty the British had totally destroyed or captured one infantry battalion, two companies of tanks, a motorised artillery battery; Combe's force took back across the Wire seven Italian officers and ninety-four other ranks, one tank and four guns. The Italian commander, Colonel D'Avanzo, had been shot dead in the fighting.

Colonel Combe won the first of his two D.S.O.s at Ghirba, the future Lieut-Colonel Seymour-Evans the second of his two M.C.s, and Lieutenant Gape of 'B' Squadron an immediate M.C.

On June 19 the burning *khamseen* began to blow and there were very bad sandstorms; the intense midsummer heat and the hot winds and sandstorms were to continue for some weeks, limiting action by both sides. Nevertheless 'A' Squadron, now operating on a huge front from southwest of Maddalena and some forty miles inside enemy territory, sent patrols to investigate El Gubi and the Giarabub Oasis eighty miles further south and 130 miles from the sea. Both proved to be strongly held and would need tanks and artillery to reduce them. The heat did not prevent the Italian air force functioning, and all squadrons of the regiment were heavily bombed and strafed in turn. Between June 19–21 however 'B' Squadron got some of their own back by destroying two planes, one a bomber, on the ground—the bomber at Sidi Azeiz aerodrome, the other in the El Gubi area. From June 20–29 the regiment was operating on a front of 120 miles—from Sidi Azeiz in the north to Giarabub in the south. It was so hot indeed that after about seven a.m. the metal of the armoured cars got too hot to touch with the bare hand, and it was certainly impossible usually to operate between 10 a.m. and 4 p.m. The sandstorms made the going often very rough, and the Rolls-Royces boiled easily and used too much water if driven during the heat of the day. And water, of course, was priceless.

On June 25 the *khamseen* blew its worst and this awful day was probably the hottest and most agonising ever experienced by the 11th in the desert. The armoured cars could hardly move because of the water shortage and the unarmoured Rhodesian scout cars trying to do their job for them were easy targets for Italian fighter-bombers. Men now really understood the agony of thirst and it was not uncommon for soldiers to collapse unconscious on the floors of their cars, just as first half, then the whole infantry platoon with 'A' Squadron had to be evacuated sick. Brigadier Clarke relates how

> 'Colonel John Combe, travelling round his scattered squadrons, was horrified to find the men becoming really frightened of their continual thirst under the piercing heat. For the first, and only, time in his life, he said, he saw a hunted look in their faces, as the *khamseen* went on blowing all day long and every day until the armoured cars became too hot to touch. For most of the time the exhausted crews lay sweating underneath them, for now one patrol alone would use more water than the daily allowance for a whole Troop.'

This was the less romantic side of the blue, the *ghot*, the desert, and the most memorable one, despite the grandeur of the star-splashed African night and the beauties of coolness and shadows on stretching sands, with only the hiss and humming of radio sets to disturb the healing silence of the dark. The Quartermastering staff and the Echelons did heroic work at this stage, and on many more occasions like it, so often in fact that it became almost taken for granted that they would have to make round trips of 100 or 150 miles, sometimes even 300, to replenish the squadrons with petrol, oil, water, ammunition, and food. All this in scorching heat. The Quartermaster himself, Captain Marshall, had to retire to hospital after superhuman efforts and was succeeded by R.S.M. Chadwick, destined to become one of the legendary figures in the desert and in the 11th Hussars for many years to come.

July 1940 was comparatively uneventful—but only comparatively. Long periods of observation were broken by brief spells of high drama, for the Italians had been reinforced and were growing more confident. Heat, flies, sand in everything, and desert sores were irritating enough, but frequent bombing and strafing were less easy to put up with. The bulk of 7th Armoured Division was withdrawn to the area south of Mersa Matruh because of the wear and tear on the vehicles, particularly on the tank tracks. At this stage Wavell and O'Connor had to go cautiously with their very limited resources. Mersa Matruh was the key to any British tactical plan now for it had a railhead and the only good water supply from Tobruk to Alexandria. Being 125 miles from the Wire it also allowed elbow room for manoeuvre. The 11th and the Support Group were therefore left to probe and patrol to find out the Italians' activities and if possible intentions, for it seemed certain that even so moderately determined an enemy would attack at some stage with such enormous superiority in numbers and equipment.

In order to find out what, if anything, the Italians were up to in the north, Lieutenant Pitman and Lance-Corporal W. Dayman were dropped off on the road between Gambut and Tobruk on the night of July 25, hid up overnight, and observed the main road next day. They were due to be picked up by another Troop of 'C' Squadron that night, but during the day this Troop was completely put out of action by nine Italian fighters. Another Troop sent to help it and collect survivors lost its Troop leader and two of its cars when also bombed and then attacked by tanks. And a third Troop sent to pick up Pitman at the appointed rendezvous—only a map reference to this Troop, of course—could find no trace of him. This was a disastrous day for 'C' Squadron, whose Troops in turn had endured about five hours of continuous bombing and machine-gunning from the air, not all of which they had been able to dodge. Eventually the search for Pitman and Dayman was given up two days later, and it was not until a few weeks later that a chivalrous gesture by the Regia Aeronautica confirmed what had happened. A dive-bomber dropped a message bag one morning over divisional headquarters containing a list of fourteen 11th Hussars who had been captured and a short medical report on each. Sergeant G. Mines, who had first dropped off Pitman and then been bombed, had died of wounds. There was also a letter from Pitman to his Colonel explaining that when they had been unable to find a Troop at the rendezvous they had started walking and had run into an Italian patrol.

The regiment, except for 'B' Squadron, was taken out of the line in August for a well-deserved rest; and for the second half of the month 'A' Squadron relieved 'B' in the Gabr Saleh area where large Italian reinforcements were seen building up. Not all suspicious activity was Italian, for T.S-M. Woodward on August 18 reported suspected minelaying 3 miles west of Pt. 212. It turned out to be a camel.

On September 13 the long expected Italian advance began, and the vanguard of Graziani's huge army rumbled cautiously over the frontier down Halfaya Pass and along the coast to Sidi Barrani, which they took on September 16. They had suffered substantially from accurate shelling by 4 R.H.A. The Support Group withdrew to the road between Siwa and Matruh, the 11th keeping contact east and south of Sidi Barrani. But thence the Italians mysteriously showed no signs of wishing to move on, although Matruh was clearly their key to success. They lingered eighty miles away from it, and continued to loiter, except for building a chain of forts between the coast and Rabia on the top of the Escarpment twenty miles inland, and erecting a monument in Barrani itself. The four divisions and their accompanying clusters of tanks were after all now sixty miles inside Egypt. But they had suffered 3,500 casualties against 150 British ones.

On September 17 the 11th were the only unit left in contact with the enemy, the Support Group having fallen back through the armoured car screen in the model of a tidy withdrawal; the infantry were ready in the Matruh defences, and both 4th and 7th Armoured Brigades were poised waiting on the Escarpment south of the town in case the Italians did decide to press on. Graziani, however, was in no hurry, and decided to stay put a month or two while his supply dumps and administrative services moved up closer.

Much patrolling and harassment went on, however, and the first of the famous 'Jock Columns' was formed, named after Lieut-Colonel J. C. Campbell, 4th R.H.A. This consisted of armoured cars for reconnaissance, 25-pounders for hitting, and infantry companies for protection and night work. The first 'Jock Column' soon began worrying the Italians in their rear areas, the 11th supplying the armoured cars. Future columns were thoroughly to establish the moral superiority already gained over the enemy. The Italians still had air superiority and were as active as ever, causing considerable casualties. Sergeant C. P. Lamb of 'C' Squadron distinguished himself first by shooting down a Breda 67 intact with a bullet through an oil pipe on September 17, then, when attacked by two fighters and twelve dive-bombers on October 8 by bringing down another one, although losing his own car to a bomb in the process. The second Breda crashed thirty yards away from Lamb and 'brewed up'. General Cadorna told John Combe after the war that the Italians at that time had no armoured cars and that brigade commanders at the front were complaining bitterly: 'What can be do? We cannot move because of these 11th Hussars!' Eventually Graziani detached a complete air squadron with the sole duty of taking care of the 11th.

On October 10 No. 2 R.A.F. Armoured Car Company—ten Rolls-Royces and some transport—was attached to the regiment and became for many weeks a temporary 'D' Squadron, doing very valuable work. This eased the shortage of cars, now decidedly serious. General Wavell, now concentrating on his own reinforcements and plans to attack the Italians off their guard, wrote in his despatch to the Secretary of State for War at the end of this first phase:

> 'The 11th Hussars, the armoured car regiment, was continuously in the front line, and usually behind that of the enemy, during the whole period; its tireless and daring search for information and constant harrying of the enemy showed a spirit and efficiency worthy of the best traditions of this fine regiment.'

For the coming British offensive 7th Armoured Division was reinforced during October and November 1940 by the 3rd Hussars with light tanks and the 2nd R.T.R. with cruisers, thus doubling its tank strength at a blow. Most important of all was the arrival in the Western Desert Force of the first 'I' tanks—fifty-seven Matildas, slow-moving and liable to breakdowns, but still better than any tank of this size previously seen. O'Connor had now 275 tanks of varying worth at his disposal, and here was to be his great advantage, for the Italians, who outnumbered him hopelessly in everything else, mustered only at first about 120. In artillery their superiority was two to one; their air force greatly outnumbered the R.A.F.; and their infantry divisions east of the Egyptian frontier alone amounted to seven divisions, with a strong armoured group. With 250 guns,

manned by the best trained and most efficient soldiers in
the Italian Army, and between 75,000–80,000 troops, they
looked on paper more than a match for two British divisions, although behind the Western Desert Force in reserve
were 4th New Zealand Division (so far only one brigade)
and 6th Australian Division. But Graziani could call on
more or less unlimited reinforcements, with his 200,000
men against 'Wavell's 30,000'—actually never more than
33,000, constantly short of transport, were engaged in the
coming battle. 4th Indian Division shared in this offensive.

The Italians helped to cut their own throats by disposing
their forces in a series of strong fortified posts which were
not mutually supporting and which large tracts of desert
separated. Further, their defences lacked depth. On the
flanks at Sidi Barrani and Sofafi, true, they were very
strong, but there were inviting gaps between their forts,
especially between Nibeiwa and Rabia. This gap had been
kept open before the offensive started, early on December 9,
by the Support Group and the 11th.

That day the 11th pushed through this gap ahead of 4th
Armoured Brigade. The task of 7th Armoured Division
was to cut off Sidi Barrani from the rear while the Indians
to the north took Nibeiwa and Tummar camps and advanced direct on the town with the 7th R.T.R. All went
splendidly and by evening of December 10 Sidi Barrani
had been recaptured; 4th Armoured Brigade had been
firmly across the main coastal road eight miles west of the
town since the previous day and with the 11th was preventing any escape from it. On December 11 7th Armoured
Brigade had to deal with Buq Buq and to prevent any withdrawal from Sofafi. It was thought that the Italians would
make a stand in front of Buq Buq, but Colonel Combe
believed that they were preparing to evacuate it, in which
case 7th Armoured Brigade's attack would be wasted.
Therefore he ordered 'C' Squadron to push out patrols
well to the west of Buq Buq early on December 11. Combe

was right, for during the night almost the whole Italian garrison had slipped away, and at 9.30 the remainder were rounded up. The squadron took over 300 prisoners in the area. It then moved on westwards, shortly to find the best part of an Italian division in position among the sand dunes and mud flats between the coast road and the sea.

Brigadier H. E. Russell, whose tank broke down, told Combe to take over his brigade and attack this force. The 8th Hussars with cruisers, two squadrons of the 3rd Hussars with light tanks and one squadron of 2nd R.T.R. with cruisers, covered by the guns of the R.H.A., successfully drove in on the Italians from the front and flank, although the 3rd Hussars, caught in a salt pan, got bogged down if they stopped moving fast and lost thirteen tanks. But victory was complete, despite the bravery of the Italian gunners, and the enemy finally broke and fled, leaving many dead behind them. Meanwhile 'C' and 'B' Squadrons of the 11th had made a detour to the west, hampered by mud-flats and very bad going, to prevent any escape to Sollum. 'B' Squadron lost four Rolls stuck in the mud but eventually managed to get two Morrises on to a track parallel with the coast road after the retreat was already under way. The moment the Italians saw one car they brought two anti-tank guns into action; just then the other car appeared and between them they killed the crews of both guns. The arrival of a third car unditched from the mud flats convinced the enemy that this was the vanguard of a larger force, for suddenly the Italians gave up. The three 11th Hussars car commanders found themselves with 1,800 prisoners, including four Colonels. Alec Halliday drove on west to cut off the Italians streaming back to their positions covering Halfaya Pass ten miles further on, and with great difficulty because of the going managed to get round the head of the column. He halted and disarmed another 2,000 Italians with three light tanks and many armed lorries and later drove back many more into the arms of Bill Cunningham. Before the end of the day these two Troops of 'B' Squadron had captured 7,000 prisoners. This staggering number was a great embarrassment, but they gave no trouble and were escorted back towards Buq Buq by a few cars and some 8th Hussars. 'C' Squadron continued the good work next day, and one Troop knocked out a field gun and two light tanks, and took another 500 prisoners, all very thirsty; one even tried to drink petrol. 'A' Squadron had been unlucky. The Italian Air Force had paid full attention to it on December 11 and disabled five armoured cars, causing several important casualties chief of whom was the squadron leader, Captain David Lloyd, M.C.

The battle of Sidi Barrani was over now, and Bardia was next target. So far the Italians had lost 14,000 prisoners (over half of whom had been taken by the regiment), sixty-eight guns and a great amount of material.

On December 13 'Combe Force', consisting of R.H.Q., 'C' and 'B' Squadrons, the 2nd R.T.R. and two batteries of 4th R.H.A. was given the job of bestriding the Bardia–Tobruk road and blocking any movement. Next day R.H.Q. had just joined 'C' in its position some twenty miles west of Bardia when they were heavily attacked from the air by Graziani's special '11th Hussars squadron'. The fighters, coming in almost at ground level and using new armour-piercing explosive bullets, caught the cars unprepared. The attack lasted over twenty minutes. Combe had been caught in the open and when he got to his feet saw nothing around him but burning vehicles and wounded men. His own car had been riddled with bullets, but his adjutant was still standing by the machine-gun he had used fearlessly throughout the raid. 'C' Squadron had fared badly; four armoured cars had been put out of action and the fitters' lorry destroyed. There were seven casualties. Later in the day there were further air attacks, which cost more casualties and another two armoured cars. But the main Italian telephone wires had been cut by the regiment.

By December 15, however, the last Italians had been driven back across the frontier, Support Group having dealt with the Sofafi area. Over 38,000 prisoners, including one Corps and four Divisional commanders, had been taken, 237 guns, seventy-three tanks, and more petrol and equipment.

But patrolling still went on, and feeling trapped and without proper communications the Italians fell back into Bardia. Not all enemy 'columns' seen turned out to be enemy; one reported by an 11th Troop leader proved on closer investigation to be camels, not for the first time.

By January 5, 1941, it was all over.

Next day Tobruk was surrounded, and on January 22 it surrendered. During all this time the regiment had been working to the west and south of the division, watching the left flank and testing out the untried desert routes that led to the Jebel Akhdar. As Brigadier Clarke writes:

> 'On the results of this reconnaissance a great deal depended. For General Wavell was now contemplating a single bold stroke which would not only open the way to Benghazi, but would also entrap in the "bulge" of the coastline which contained the Jebel Akhdar every Italian soldier left in Cyrenaica. To accomplish this he intended to send the 7th Armoured Division by itself into the southern desert to drive straight to the Gulf of Sirte, unsupported over a waterless 200 miles, and there to bar the coast road that ran southward from Benghazi. His infantry, meanwhile, would force back through the Jebel the Italian main body in the north.'

On January 16 Lieutenant Friend's troop ran into a mobile column near Mechili and was nearly captured. He then found that infantry accompanied by tanks were digging trenches at the entrance to a pass beyond Mechili which controlled further passage along the Trigh Enver Bey. The 11th were therefore left out of the Tobruk attack—some seventy miles away—to keep observation on Mechili, and

Imperial War Museum

In the wake of the retreating Italian Army discarded helmets litter the desert.

some patrols reached Bomba and the outskirts of Derna before Tobruk fell.

On January 24 4th Armoured Brigade fought a sharp action at Mechili, but although they had a very good position the Italians withdrew by night into the Jebel Akhdar, followed by the 11th and Support Group. 7th Armoured Division was now down to under seventy cruiser tanks.

Derna fell to the Australians on January 30, and on February 2 'C' Squadron made contact with the left flank of the Australians. Now 'B' came under Australian command for the drive along the coast to Benghazi, while the rest of the regiment, brought up to strength by a squadron of the King's Dragoon Guards newly arrived in Africa, rejoined their division for the final advance across the desert.

After Wavell had flown up to consult O'Connor and Creagh it was confirmed that the Italians must be cut off south of Benghazi. The advance, led by the 11th and 4th Armoured Brigade, started from Mechili early on February 4, the division now having fifty cruisers and eighty light tanks. Fort Msus was soon captured, and 'C' Squadron reached the outskirts of Antelat, another thirty miles on, by evening. 7th Armoured Division was concentrating and replenishing round Msus, expecting to continue due west, when news came that the enemy had already begun withdrawing from Benghazi. General Creagh decided to depart from his original instructions in the light of this, and to strike south-west immediately as fast as possible to trap the Italians. General O'Connor at once approved this decision when it was communicated to him. During the night a mobile flying column was put together with Colonel Combe in command, including the 11th Hussars.

Early on February 5 this column started off after 'C' Squadron towards Antelat, with the promise that 4th Armoured Brigade's tanks would follow as soon as possible. At first the going was exceptionally bad, and the wheeled vehicles of the Rifle Brigade and the artillery, which had been ordered to join Colonel Combe far in front (the 11th having managed to cross the worst of the going), fared badly. Tracked vehicles like tanks and Bren carriers soon caught up with some of the wheeled vehicles that had started in front of them. 'This stretch of bad going lasted for about twenty miles and must be one of the longest stretches of almost impassable country that a mechanised force has ever been ordered to cross,' General Verney states.* When the going improved the wheeled column quickly drew ahead,

* In *The Desert Rats*, Hutchinson, 1954.

11 H.—9

but it was very hot and the going made the Bren carriers of the Rifle Brigade use so much fuel that they ran out of petrol and could take no part in the battle. There were also unexploded thermos bombs to dodge and attacks from the Italian air force. Despite these and a complete lack of maps the column made good time to Antelat, only to find that 'C' Squadron had already got patrols up to the coast road near Sidi Saleh, but, fortunately, had so far seen no enemy. Shortly after noon Colonel Combe had his whole force in position right across the last line of retreat for the Italians. The K.D.G. squadron was put to the south to give warning of any attack from Tripolitania, met some opposition, and took many prisoners.

There was not long to wait. At 2.30 p.m. the first of a large column of Italians, the retreating Benghazi garrison, came into view. The whole column was estimated at 20,000 men with tanks and guns. Combe had less than 2,000 men with twenty guns and no tanks, although the 4th Armoured Brigade, it was hoped, would join him in three hours. His own 'B' Squadron was still with the Australians, 'A' and 'D' with the Support Group, still thirty-five miles away to the north, but pressing on to the coast road at Ghemines to take the enemy in the rear.

The first shots of the battle of Beda Fomm were fired by 'C' Squadron and a Rifle Brigade Company, who met the Italian vanguard head on. Then the R.H.A. joined in, while the rest of the R.B.s moved into positions to defend the guns. The Italians were so demoralised by fear and surprise, that they did not think of taking things calmly and making a carefully thought out attack on the small British force, outflanking it for a start, but put in a series of frontal attacks which the R.B.s beat off, while 'C' Squadron looked after the right flank. Then they did have some success infiltrating by way of the seashore, so the R.B.s had to extend their line.

At 5.30 p.m. 'Combe Force' was heartily relieved to see the first tanks of 4th Armoured Brigade driving towards

Beda Fomm, towards which they had been diverted at Combe's suggestion so as to take the Italian column in its left flank and rear. Beda Fomm was found clear and the 7th Hussars immediately attacked, causing panic among the Italians and setting fire to petrol lorries. They took over 1,000 prisoners before darkness fell. 4th R.H.A. at the other end were now down to thirty rounds per gun.

The Italians were now caught in a flattish plain between Beda Fomm and Sidi Saleh along fourteen miles odd of road with sand dunes and the Mediterranean on their right and the British—in unknown strength to them—on their left, and at each end. Skilful British patrols that night contrived to give the impression of a much larger force's presence. The night was fairly quiet. Next morning, February 6, the main Benghazi garrison began converging upon Beda Fomm and there was fierce fighting throughout the day. The Italians could not break out at either end. They were kept bottled in at the Beda Fomm end; at the Combe Force end, where the Rifle Brigade, 4th R.H.A. and 'C' Squadron and 'R.H.Q.' were covering three and a half miles of defensive position, every attack down the main road or to its flank was beaten back by the British. All day long both forces held out, taking more and more prisoners. With the Australians now in Benghazi 'B' Squadron was released to join the Support Group, and 1st R.T.R. were haring up from Msus to reinforce the tanks. The cruiser strength of 4th Armoured Brigade was down to nine, had the Italians, who easily outnumbered our tanks, but known it. At last a large number of Italian tanks supported by infantry did break through the 2nd Tanks, by now down to seven cruisers, towards nightfall, but Combe was instantly warned and they were received with vigour by his force. Only four got through to safety.

On the night of February 6–7 came the most dangerous attack. In desperation the Italians moved a number of tanks heavily supported by infantry past the seaward flank of the

Rifle Brigade by the sand dunes. They made nine consecutive attacks to break the Rifle Brigade's hold on the coast road, but although one reached the reserve company the thin green line never broke. The night, like its predecessor, was very cold, as it often was in the desert, and very wet. The Rifle Brigade, greatly outnumbered in exposed positions, were magnificently backed up by 4th R.H.A. and the battery of 106th R.H.A.

Soon after dawn on February 7 the Italians made their final fling and attacked with infantry and thirty medium tanks, but still the Greenjackets and the gunners, not to mention the 11th, stood fast and beat them back, though

Imperial War Museum
Captured Italian tanks.

ammunition was fast running out. It had become too much, and the whole Italian force suddenly surrendered. By 8.30 a.m. the battle of Beda Fomm was over and the Italian 10th Army no longer existed. General Tellera, its commander, had been killed, and General Bergonzoli, known as 'Electric Whiskers', a renowned figure in Libya, made the surrender to Major Pearson of the Rifle Brigade, and then to John Combe. Five other generals, including General Babini who had commanded 160 tanks, also surrendered, and another 8,000 prisoners. Bergonzoli, who had made a sensational escape from Bardia, told Combe he had been amazed at encountering the 11th Hussars across the main road on February 5, as the Italian Staff thought the British were still two days' march away and heading directly for Benghazi.

'C' Squadron still pushed on, and at noon entered Agedabya with two Troops and took 200 prisoners. During the next few days the rest of the regiment came up and pushed on to El Agheila, later sending patrols into Tripolitania as far as Marble Arch.

But already O'Connor had radioed Wavell in Cairo—in clear, in case Mussolini should be listening in—'Fox killed in open.' The astonishing bloody battlefield of Beda Fomm, littered with knocked-out tanks, lorries, guns and war debris, the main road blocked and cluttered by vehicles and more guns, signalled itself eloquently enough a complete and tremendous British victory.

The final act in the battle for Cyrenaica had cost the Italians at Sidi Saleh and Beda Fomm over 20,000 men, 216 guns, 112 tanks. It brought their total casualties to more than 130,000 men, 400 tanks, and 1,290 guns in two months; during that same period the Army of the Nile had advanced more than 500 miles and destroyed ten divisions, at a cost in total casualties to itself of under 2,000. This had been done with two divisions of which the original 'Desert Rats', the 7th Armoured, was one.

It was indeed John Combe's finest hour.

Chapter 13
Rommel: The Strongest Enemy

'A week was seven domes across a desert,
And any afternoon took long to die.'

(Mark Van Doren.)

OPINIONS vary as to whether, after Beda Fomm, the British should have continued their advance into Tripolitania. The lines of communication had been stretched so far, transport, even with the welcome addition of hundreds of captured lorries, was getting so worn out, that a further advance would have been something of a gamble, even if it was an exaggeration to say, in General Collins's words, that 'In Cyrenaica XIII Corps had now reached the end of its administrative tether,' and that 'no further advance was possible until Benghazi could be developed as a port.' Yet the Italians were completely demoralised, and had little with which to stop a new attack. O'Connor thought differently from his masters in London, and Wavell, though full of doubts already about the Balkans and Greece, backed him up. Both realised that full support from the Royal Navy and the R.A.F. would be necessary, but O'Connor envisaged a swift attack on Sirte by the Support Group, the 11th and the R.H.A., followed by a decisive blow against Tripoli by 7th Armoured Division's remaining tanks augmented by some from the newly arrived 2nd Armoured Division in Egypt. He also envisaged the landing by sea of an infantry brigade at Tripoli. It was above all a case of 'If 'twere done, 'twere best done quickly.'

General Erwin Rommel, who had just been entrusted by Hitler with command of the Afrika Korps being sent to help the Italians, agreed with O'Connor. He wrote in February 1941:

> 'If Wavell had now continued his advance into Tripolitania, no resistance worthy of the name could have been mounted against him—so well had his superbly planned offensive succeeded.'

And, typically, he added:

> 'When a commander has won a decisive victory— and Wavell's victory over the Italians was devastating —it is generally wrong for him to be satisfied with too narrow a strategic aim. For that is the time to exploit success.'

Thus, when Rommel arrived in Sicily on February 11, he was quite prepared to hear in the next few days that British troops had reached the outskirts of Tripoli, and with Hitler's approval ordered General Geissler's X Luftwaffe Korps to 'cripple the British supply traffic to Benghazi.' Actually Rommel landed at Tripoli two days later.

In the desert there had long been rumours about the German Air Force, supposedly supplying pilots to fly Italian planes, and it was no great surprise when on February 14 near Agheila 'C' Squadron was attacked first by one, then by two squadrons of Messerschmitt 110s, which put out of action an armoured car and the fitters' lorry. Later that day the Fort was dive-bombed by Junkers 87s and another car and two lorries were hit. This was the first time that anyone in the desert had been exposed to the fearsome Stukas, with their shrieking noise as they dived looking like great crouched birds. Another heavy air attack was made on Fort Agheila next day. Clearly the Luftwaffe meant business.

As a result of the high level deliberations in London and Cairo most of 7th Armoured Division was now withdrawn to Egypt to refit and rest—much needed occupations, it must be admitted—handing over to one brigade of the newcomers, 2nd Armoured Division (whose other brigade, unfortunately, was to go to Greece in April). The 11th, sadly saying goodbye to their good friends from the R.A.F. Armoured Car Company, handed over to the King's Dragoon Guards, and on February 22 reached Abassia—out of the line for the first time as a regiment after nine months' fighting.

Wavell, badly short of good intelligence across the Tripolitanian frontier, did not consider the Germans likely to try to recover Benghazi on March 2, since by then only one Panzer regiment had reached Africa. Nor did he and others envisage any large-scale German operations much before mid-April. But nor did Hitler. Both under-estimated Rommel's intuition and dash. In fact Rommel attacked to re-take Cyrenaica on March 31, with only the German 5th Light Division and two good Italian divisions, the Ariete and the Brescia. By this time, Wavell, still having to look in several different directions at once, was somewhat worried. For O'Connor, who had been unwell, had handed over command to Lieut-General P. Neame, V.C., who had no experience of desert warfare; nor had Major-General Gambier-Parry, with his ill-equipped half an armoured division, some of its tanks being obsolete, others captured Italian ones; the new Australians, though likely to be splendid troops, *were* new; and transport was shockingly deficient, 8,000 vehicles having gone to Greece and the promised American Lease-Lend trucks only beginning to arrive.

All these circumstances paved the way for the defeats of 1941–42 and the long, slogging, to-ing and fro-ing over the next eighteen months, the bloody and wearisome set-piece and impromptu battles, the abortive 'Battleaxe', the more successful 'Crusader', the endlessly wearisome desert war—at the end of which the British stood with their backs really to the wall at Alam Halfa and Alamein before the final great advance to victory. Rommel, like his opponents, was often subject to pressures and interference from home, but was able better to withstand them, and remained in sole effective charge until near the end. The British, however, with their masterly knack of taking it out of their senior commanders at inopportune moments, almost deserved to lose sometimes, when one thinks of the chivvying and interference, bullying, and maladroit misunderstanding that men such as Wavell, Cunningham, Auchinleck, had to put up with. For that Churchill, great inspirer though he was, must bear a full share of the blame, and to a lesser extent the War Office. It was not until the admirable partnership of Montgomery with and under Alexander that full confidence was re-established.

★　★　★　★　★　★

Back in Cairo in March the 11th lost the last of their Rolls-Royces, which were replaced by Marmon-Harringtons carrying a crew of four. But these never turned out very happily, with their inferior armour and springing, and were only to last a few months.

It is worth noting at this stage that throughout the war the 11th Hussars was not an easy regiment to get into, for either officers or, to some extent, men. This, of course, applied to other regiments, although armoured troops' requirements were always especially high. But once in it, one had to prove one's worth to stay in it. Some soldiers, both officers and men, were more suited to fighting or jobs other than 'leading the advance in tin boxes on wheels with pea-shooters', as one infantry colonel once put it, with the chance of an 88 mm. or, later, a bazooka over the next ridge or at the end of the agonisingly straight road or in a ditch round the next corner. This was often no reflection on them, and they did sterling work elsewhere or in another capacity. But they had to be sorted out. On the other hand, there were 11th Hussars of all kinds throughout the war who simply would not be separated for too long from the regiment, once they had recovered from wounds, illness, training other people, or act-of-God postings or divorcement. Lance-Corporal R. H. Garner, D.C.M., for instance after years of 'stone-breaking' in Italian and German camps, rejoined the regiment towards the end of the war and in the space of several weeks shot more Germans than fell to the lot of most people during the whole war. There were too those who, overdue for return to comparative tranquillity in England with honour after long service, simply refused to go. And of the fifteen or twenty 11th Hussar N.C.O.s who were commissioned from the regiment, nearly half returned to fight with it.

In this general context it is not amiss to quote from *Tail of an Army*, by Lieut-Colonel J. K. Stanford, who travelled out with Major A. T. Smail's draft to Egypt in 1941. For the view of a regiment seen through outside eyes is always interesting, often clearer than that from within it.

The ship in which Stanford travelled contained a hotchpotch of units and drafts, including what he called 'a mob destined for "G.H.Q. 2nd Echelon"', but also some infantry and other arms.

> 'Most colourful, and with a cachet of their own, was the little bevy of officers of the 11th Hussars. Their senior major, Trevor Smail, an ex-master of foxhounds, was visibly preparing himself for the travail of the desert, which meant for him command of his famous regiment and a double D.S.O. The other major, Ken Alexander, had been on Lord Rawlinson's staff in India.
>
> 'Normally these Hussars sat about in their cherry-coloured trousers, looking beautiful and aloof, but their aloofness hid much quiet efficiency. They did not join the lamentations about the filth of their quarters and the messing utensils. One day they invited me to join an inter-squadron competition which they and the Tank Regiment were holding. We were plagued with flies picked up at Mombasa, our tables grimy and our mess

equipment rusted. Other detachments declined to improve things. They merely groused. But these two units had set to work, and their burnished pots and pans, their scrubbed and shiny tables, showed what regulars of long tradition could achieve.

'Their inspection shamed me. It was my first introduction to crack cavalry since the Villers Bretonneux battle of 1918.'

By April 3, 1941, the Australians had been ordered to evacuate Benghazi. On April 8 the emasculated 2nd Armoured Division, which already had suffered a series of setbacks, met final disaster at Mechili when General Neame was trying to hold a defensive line between there and Derna. There was then only 9th Australian Division and one brigade of 7th Australian Division to man Tobruk's western defences and stop the enemy.

On April 3 John Combe had been promoted Brigadier, and was due to go to Palestine, so next morning Lieut-Colonel W. I. Leetham took command of the regiment.

Combe never reached Palestine, for a few days later, having been summoned as a desert adviser to Neame at O'Connor's request, all three were captured by an advanced German column as Neame, advised by O'Connor, was directing the retreat from Derna to Tobruk. O'Connor and Combe were to escape from Italy later in the war, and the former commanded a Corps with distinction in Normandy, the latter an armoured brigade in Italy. But for their years in prison camps (and at large in Italy for months) they would surely both have reached even greater heights than they did.

On April 5 the 11th rushed back to the desert, less 'B' Squadron, still waiting to be re-equipped, which rejoined the regiment at the end of the month. On April 8 at El Adem Brigadier 'Strafer' Gott, now commanding the Support Group, ordered Colonel Leetham to patrol a fifty-mile front from Acroma west of Tobruk to Bir Hachiem and Bir Gobi, and to delay the enemy as much as possible while the Tobruk defences were being prepared. These tasks were carried out until Tobruk was surrounded and the Support Group, fighting all the way, eventually halted in a good defensive position near Sollum behind the frontier. Tobruk was to hold out for seven months more, a feat which ultimately undid the Afrika Korps. The Germans and Italians stopped at the frontier, up against supply problems themselves, and if they could have used Tobruk for their ships things might have turned out very differently.

In May an attempt was made to relieve Tobruk, with far too few tanks, but achieved nothing more than the recapture of Halfaya and Capuzzo, and at the end of May the Germans attacked and recaptured Halfaya. The war was becoming fluid but full of surprises. Thus for 'A' Squadron May 26 was a very quiet day until mid-afternoon—when an enemy column of about 200 vehicles, including fifty tanks, came through the Wire and reached Hamra. Lieutenant M. M. C. Clark, M.C., made contact with this column, and kept it under observation, his Troop being shelled for two and a half hours with one car hit.

11th Hussars had started the war in an armoured brigade, but it had very soon become obvious that they must act as divisional troops (often a division needed two armoured car regiments in the desert) and this they had now been for some time and were normally to remain, except when working directly under a Corps, for the rest of the war.

Operation 'Battleaxe' between June 15–18 was designed to relieve Tobruk. It failed in this objective, but at least prevented the Germans from mounting any big new advance. Rommel's 5th Light Armoured Division had now been reinforced by the more powerful 15th Panzer Division. The British used 4th Indian Division, now back, 22nd Guards Brigade, and 7th Armoured Division.

Lieut-Colonel Carver wrote in 1943:

> 'This was the first occasion when the Division engaged a German Armoured Division equipped with

Mark III tanks and 88 mm. guns, in a full-scale tank battle. It ended in failure, 6 Battalion R.T.R. suffering heavily at Hafid ridge, and many Matildas, which were employed in the role of cruiser tanks, breaking down before they were knocked out. Support Group gallantly held one German counter-attack, 4 Regiment R.H.A. under Lieut-Colonel Campbell using their 25-pounders with great effect. After "Battleaxe", Support Group and 11 Hussars were left in the area of Sofafi, with Guards Brigade under their command at Buq Buq, while the rest of the Division withdrew to the area of Matruh, 4 Armoured Brigade returning to the Delta to re-equip.'

Both sides over-estimated not only each other's tank strength before the battle but their losses during it. Rommel claimed 220 British tanks lost, whereas actually fifty-eight 'I' tanks and twenty-nine cruisers were destroyed or captured. The British claimed nearly 100 German tanks knocked out, Rommel admitted only twenty-five completely destroyed. The truth probably lay in between.

On the first day of the offensive 'A' Squadron, now commanded by Captain John Lawson, M.C., lost a complete Troop and had seven casualties when seven Messerschmitt 110s attacked them. Six days earlier the squadron had picked up, near Sidi Barrani, 139 British and Commonwealth officers and men who had spent a week in an open boat crossing from Crete—an ominous piece of news.

The R.A.F. now had a number of Hurricanes, but these were sometimes a double-edged weapon as far as the regiment was concerned, though invaluable generally. Several times forward troops or even squadron headquarters were attacked by them. The Italians still kept busy in the air, and on July 13 Sergeant E. D. McCarthy, M.M., of 'B' Squadron lost two armoured cars when attacked by six Macchi fighters.

Colonel Leetham summed up of the summer of 1941:

'And so the summer passed with armoured cars continuously either in touch with or reporting on or harassing or watching the enemy day and night. Incidents of individual hardship and bravery occurred weekly; of troops being put out on patrol for ninety-eight hours on end—short of water and fresh rations—observing through all the long days of light amidst a multitude of dive-bombing, flies and, further, during the hours of darkness never certain whether an attack was imminent, which meant no proper sleep and all the men on edge. Of armoured cars far behind the enemy front line being totally disabled by enemy aircraft and crews crawling to cover to avoid being machine-gunned from the air; burning their maps and guns and walking across the desert towards our own lines—sometimes reaching them, sometimes being picked up by reinforcement patrols and sometimes never heard of again.

The regiment had been pleased to have a squadron of the Royals attached to it in July, and it had been intended that when the Royals were fully trained they would take over as a regiment from the 11th; this in fact had all but taken place later that month when the Royals were transferred to Palestine except for one squadron. However Lieut-Colonel Newton King and officers of the 4th South African Armoured Car Regiment, fresh from successes in Abyssinia, had also been attached to the regiment since early June, and from mid-August this good regiment gradually took over. In both cases the recipients freely acknowledged and welcomed the help given them by the 11th Hussars.

When the regiment at last returned to Cairo in mid-September General Wavell had gone, replaced by General Sir Claude Auchinleck. Fittingly, the troops had last seen him when, worried by tank losses, he visited the battlefield in a light plane during 'Battleaxe', and confirmed the decision to withdraw—a withdrawal made possible without disaster only because of the incessant and accurate bombing by the R.A.F. of the German Panzers.

In Cairo the regiment rested and refitted again. Had Italian newspapers been available it might have thanked its lucky stars it was still functioning on a comparatively modest basis, for on June 18 at the end of 'Battleaxe' United Press had quoted Virginio Gayda, a Fascist editor, as saying that British forces in the Sollum area 'included 1,000 tanks, comprising two British Divisions, one of which was the famous 11th Hussar division'.

Rommel was busy reorganising too. The 5th Light was converted into the 21st Panzer Division, and the 90th Light Division, to become known as a formidable opponent, arrived during the summer.

Brigadier Gott wrote to Colonel Leetham:

> 'The respite you provided has enabled the Army of the Nile to become once again strong, well-trained and ready. You can look back on the summer of 1941 as a time when you gave valuable assistance towards our ultimate Victory.'

Rommel had handed over command of the Afrika Korps in August to Lieut-General Cruewell, since his own command was now a Panzer Group; he also had under him six Italian divisions, two of which were motorised and had some tanks. Rommel was having troubles of his own, the chief being that Hitler and the O.K.W. in Berlin seemed inclined to regard North Africa as a sideshow and Rommel's duty simply as rescuing the Italians. This did not sit well with the Afrika Korps. In addition, since the early autumn if not before the strengthened Desert Air Force of the R.A.F. had established definite air superiority over the Luftwaffe and Regia Aeronautica. Although the bombers, particularly the diving Stukas, of the latter were dangerous and feared, and the Messerschmitts were equally dangerous to forward troops, the Luftwaffe was desperately short of fighter squadrons, although Rommel and General Froelich had repeatedly asked for more. Everyone had been delighted when the R.A.F. had bombed a headquarters in May 1941 which the Inspector-General of the Luftwaffe, Field-Marshal Milch, had been visiting. Milch, in a spotless white uniform, had had to take cover by diving into a trench filled with kitchen garbage.

In tanks, however, Rommel still had superiority, not of numbers but of efficiency and firepower. It might be held by pundits that the 2-pounder at 1,000 or more likely 800 yards had greater penetration than the 50 mm. or short 75 mm. of the Germans, but the German tanks, of course, could stand back and pick off the British ones at ranges greater than those. And with the long-barrelled 75s that the later Mark IVs had from 1942 this superiority was even more decisive. It accounts for the great German successes in tank battles up to the time in May 1942 when the British began to get American Grants and Lees, with 75 mm. guns albeit a limited traverse, and later the much more efficient Shermans, and eventually to put some 6-pounders on their own British cruiser tanks. Rommel himself had written after 'Battleaxe' of the new Crusader (still with only a 2-pounder): 'Had this tank been equipped with a heavier gun, it could have made things extremely unpleasant for us.' Most British anti-tank guns were still only 2-pounders also.

Since June 1941 Rommel's six- and eight-wheeled reconnaissance armoured cars were constantly seen, and often engaged by the 11th Hussars. They too were a great advance on anything that the Axis had been able to provide for their purpose previously. Indeed on one occasion on August 7 two of them had penetrated 'A' Squadron's lines early in the morning, either undetected or mistaken for Marmon-Harringtons, attacked a Troop at great speed and knocked out two cars killing four men, and then made off equally fast taking with them a sergeant and three men of the 11th. This in its way was a copybook armoured car operation, and was helped by a spotter plane overhead.

In the second week of October new Humber armoured cars began to be issued, but had to have a number of ordnance modifications. Next door the 8th Hussars, old friends, were gladly giving up obsolete light tanks in exchange for American 'Honeys'. The Marmon-Harrington had been unpopular with the troops, and no match for its German opponents. But the Humber was a very different matter. It had two fine machine-guns, the 15 mm. and 7·92 Besa, and if the former was rather subject to stoppages the latter was one of the best guns of the war. The Humber's armour was good, and it was fast, though its engine was underpowered and had too short a life of only 3,000 miles. By the second week in November the whole regiment was back in the desert with their Humbers—the best tools they had been given so far.

Another new development had been the creation of the Eighth Army, now consisting of two Corps, XIII, and XXX under an old 11th Hussar, Lieut-General Willoughby Norrie, D.S.O., M.C. The regiment and 7th Armoured Division were now in the latter.

Rommel had been agitating for some time to launch a new attack against Tobruk, with only discouragement from his superiors in Germany. But he eventually received permission, and set November 23 as the day for his attack. He was forestalled by 'Operation Crusader' which Auchinleck launched early on November 18 while Rommel was still taking a brief holiday in Rome.

The desert was again becoming more crowded, and 7th Armoured Division, now commanded by Gott, besides its usual brigades and the Support Group, had more divisional troops and a complete new armoured brigade. In tank *numbers* the division was more than a match for Rommel's two armoured divisions, but instead of operating in mixed groups of various arms there was too sharp a dividing line between armour and infantry drawn by Cunningham.

In the opening stages of the battle on November 18–19

the regiment was in action against the Italian Ariete Division with its considerable armour, in the El Gubi area on the general line of the Trigh El Abd. Many running fights were fought, and Lieutenant P. F. Stewart of 'C' Squadron gave an account of a typical day which has never been bettered. Stewart's troop had been sent to help another in trouble with jammed guns, when both were shelled accurately:

> 'The ground was like a pancake, but the only thing to do seemed to be to move, so we moved. Suddenly saw about eleven vehicles in front, and till they opened up thought they were ours.
>
> 'Moving fast on dead flat country, shouting instructions to the driver, traversing, firing, clearing stoppages on the Besas, reporting, perhaps with one foot through a loop in the microphone flex or a bit of tobacco ash in one's eye, with the floor of the car slippery with empty shell-cases and the turret filled with the sharp smell of cordite, one had no time for being frightened or to listen to the muffled "Wham! Wham!" of shells passing overhead. One had no time for navigation either. It was hard enough, with the turret traversed and giving the impression that the car was moving sideways, to go where one wished, let alone know where it was. The ground and the position of the enemy were the dictators of movement, not one's map or one's compass.
>
> 'This time I got some good ground and fought quite a rousing little fight till they withdrew. These actions seldom come to anything. For a few overcrowded minutes all life was concentrated in the brief fury of noise and smell and confusion—and then it was all over. One or the other pulled out of range or became lost behind a ridge of sand, and existence slowed down to the old routine. One "brewed up", looked for bullet holes, swept out the empty cases and cleaned the guns. One settled down once more to the familiar business of watching, reporting, being shelled and being bored. One rather hoped it would happen again.

> 'Here the desert last year was dry and flat in the blazing heat. Now heavy rain clouds massing by the declining sun and more and more pools of lying water. Host of stuff moving parallel with me, looking very like Germans, but of course not.
>
> 'Last light and car stuck in the mud and another pulling it out. Suddenly my heart in the pit of my stomach, sick with fear, seeing a tank 300 yards away with its guns on me. Ours. Uneasy night after cold bully and sauce which was ambrosial.' [*The Eleventh At War*.]

The situation was soon to become even more obscure, especially as wireless contact was frequently lost. The great and bloody tank and infantry battles round Sidi Rezegh that followed remain in some parts obscure even today, although the tank casualties were clear enough. By November 25 7th Armoured Brigade had ceased to exist and its remnants were sent back to base, never to rejoin the Division. By the end of the 23rd the 5th South African Brigade, which, with the rear of 7th Armoured Division and the head of 1st South African Brigade, had run unexpectedly into 15th Panzer Division attacking, had also more or less exhausted itself as a fighting formation. The first phase of Sidi Rezegh from November 21–23, was a decided German victory.

Lieut-General Fritz Bayerlein, at the time Chief of Staff to Cruewell, and later to Rommel, wrote later:

> 'The British obliged by throwing their armoured brigades into the battle in separate units. This enabled us to gain a series of partial successes, and eventually led to victory in one of the greatest armoured battles of the campaign, in which the bulk of the enemy's armour was destroyed.'

4th and 22nd Armoured Brigades had also lost heavily in tanks—indeed by the end of the 23rd two-thirds of the British armour was out of action. The rest and scores of other vehicles were dispersed in confusion, and General

Verney has described how Brigadier Jock Campbell, who had just won the V.C. directing the Support Group from his open staff car, with what was left of *it*, ' "C " Squadron of the 11th Hussars and a dozen tanks of the 7th Hussars, performed tremendous feats in rallying the scattered drivers.' The 11th lost more men killed, wounded and captured on the 23rd than on any previous day before. All was confusion. Each squadron was separated by miles, sometimes each troop, and R.H.Q. was as much in the battle as anyone. Squadrons and Troops faced all four points of the compass at different times, going towards the sound of gunfire wherever the Germans appeared, sometimes fighting under the nearest formation of friends they could find, sometimes giving their own orders. Stewart wrote:

> 'I do not know what happened that day. I don't think anyone does. . . . Not only was one very wrapped up in one's own personal little battles, but one had no information of what was going on elsewhere. We fought privately and bitterly.' [*Ibid.*]

Rommel's losses had also been heavy, not only in infantry, for by the evening of November 23 he was down to about 100 German tanks out of his original 260. (The Italians had 154.) The British had lost more than twice as many tanks as had Rommel. This was very largely due to the hopelessly inadequate 2-pounder and the unstoppable German 88 mms.

But it was also due to the inspired handling of the German armour by General Cruewell and to Rommel's leadership, in the face of paper odds in tanks of almost two to one. More than two to one if British tank reserves, sent up at the rate of about fifty a day, were included. Besides, the British had great air superiority. The British infantry had been placed in pitiful situations with nothing to stop German armour, for 2-pounder anti-tank guns just as those in tanks bounced off the German tanks at longer ranges.

Those at home to blame for this appalling armaments situation at that time in North Africa should have their names inscribed in blood on a stone memorial in the desert.

The British defeat had very nearly become a rout. The South Africans and 22nd Armoured Brigade and the Support Group had done much to stop it, although on November 25 the German radio announced that '22 British Panzer Division was entirely surrounded and being mopped up'. In this second phase of the battle of Sidi Rezegh, when Rommel had decided to make an all-out thrust for the frontier and the sea, confusion still dominated much of the battlefield, as both sides recovered tanks that could still run. 'A' Squadron eventually managed to link up with R.H.Q. after having been heavily engaged by South African anti-tank guns and 25-pounders, and 'B' Squadron found themselves competing with Germans and Italians in 22nd Armoured Brigade's area as they salvaged tanks. The 11th reported the recovery of seventy tanks during the day.

Rommel's advance was touch and go, and hung in the balance the whole of November 25 and 26, but the great resistance of the New Zealand Division in the north and the 4th Indian Division at Sidi Omar further south played a large part in stopping him. The Indians reduced the tank strength of 21st Panzer Division to ten tanks. 7th Armoured continued to fight fiercely, though by November 29 22nd Armoured Brigade (Brigadier J. Scott-Cockburn) had only twenty tanks left commanded by Major Kidston. That evening they were relieved by 4th Armoured Brigade; the 11th were sorry to see them go, as during their short association 'we had all got on most wonderfully well together'. The association was to be renewed before long.

By the end of November Rommel had nearly shot his bolt, although again in possession of Sidi Rezegh, which had changed hands twice, and although he had had spectacular successes, such as the destruction of most of the New

Zealand Division which had linked up with the Tobruk garrison. But he could not take Tobruk itself and was being heavily menaced by British troops in his rear. His casualties had also been heavy. Ironically, had he turned south instead of north after his initial success, to mop up XXX Corps, he would almost certainly have captured the two huge field supply dumps upon which the whole British force depended, each some fifteen miles away from Gabr Saleh, and at that point able to call upon only 22nd Guards Brigade to defend them. As it was the Afrika Korps actually passed through the water point on the northern edge of one of these dumps without recognising it for what it was.

There was more fierce fighting in December, but by the 6th it was obvious that Rommel was going to withdraw. By December 10 the Germans had been driven clear of the Tobruk perimeter, and struggled back fighting hard to the Gazala line. Here they stood and made one savage counter-attack against 4th Indian Division, but the 4th Armoured Brigade turned Rommel's southern flank and he was forced to withdraw on the 16th to Mechili, and thence helter-skelter right back to Agedabya, followed by the remnants of 7th Armoured Division. In the process Rommel lost half his few remaining tanks and left large numbers of Italians stranded without transport. Finally at the end of December 1941 he retired still further to the greater safety of the El Agheila area and the bulk of the Axis forces concentrated in Tripolitania to lick their wounds. The end of 'Crusader' was a British victory, after the scarifying early tank defeats.

Rommel, however, had promised his own troops and the natives of Cyrenaica that he would shortly be back, and he was to be as good as his word. For the time being it was essential for him to repair his losses, in the end much larger than those of the British. Even in armour, because of his retreat which enabled the British to recover many of their own lost or abandoned tanks, his permanent loss in German and Italian tanks, about 300, was greater than the British—278.

The 11th had fought hard and they had fought all over the place. 'B' Squadron, for example, had been completely absent for three weeks at one stage attached to tanks. Nearly every Troop in the regiment had had its share of adventures, some extraordinary. It was frequently extremely hard to tell enemy vehicles from friends; one day a lost 'B' Squadron lorry joined for 'protection' a German column of trucks to avoid the hot pursuit of what transpired to be R.H.Q. of the K.D.G.s; a Troop leader with a large party of 22nd Armoured Brigade being chased by fifty German tanks passed within 200 yards of an enemy anti-tank position but dissuaded them from firing by cheerfully waving his beret at the gunners; Sergeant C. P. Lamb, D.C.M., M.M., of 'C' Squadron was missing with the squadron's wireless Humber for eight days, during which he supplied a valuable wireless link with one of the mobile columns; Lance-Corporal H. P. Lyon of 'B' Squadron, whose car had no wireless, recovered a missing Humber but then had to wander, often fighting, about the desert for ten days before he found his fellows again; 1 Troop, 'A' Squadron captured first a German staff car with brand-new codes, and later one of von Ravenstein's A.C.V.s containing important papers and maps; 2 Troop destroyed 15,000 gallons of aviation petrol on Gasr el Aird aerodrome; Sergeant Peacock of 'C' Squadron destroyed two Junkers 78s on the ground and the next day eight anti-tank guns; the regiment killed many Germans, took nearly 400 prisoners, knocked out or captured several light tanks, some fifty lorries, and a few more anti-tank and field guns. Its own casualties had been remarkably light.

In the New Year, ushered in by appalling weather including both sand- and rain-storms, 7th Armoured Division went back to Cairo by stages, being relieved by the newly arrived 1st Armoured Division. Only the 11th stayed through January at the front, with the Guards Brigade, Support Group having left for Egypt on January 18. They operated in the Antelat–Msus–Agedabya area. 1942 began inauspiciously. In the Agedabya area minefields gave a lot of trouble, and 'A' Squadron, working under Colonel Mowbray's Coldstream Guards column lost four armoured cars and three lorries blown up. The going was also terrible in many places, and vehicle casualties became so bad that after 'B' Squadron, which had suffered worst during the last two months, was brought up to strength in cars 'C' and 'A' had to be merged temporarily into a composite squadron with Lawson in command and Captain W. V. Burdon, M.C., assisting him.

New subalterns joining the regiment at this time soon found that they were not allowed to take over a single car, let alone a Troop, until they had been thoroughly trained in every aspect of the Troop's work. Disconcerting although this practice—which the 11th had been the first to introduce —may have been to some who had been 'leading' Troops in England for months or even years, the scourge of farmers and the heroes of many an unrealistic exercise, it undoubtedly saved countless lives. It had become absolutely axiomatic in the regiment, and was to remain so until the end of the war. It was not considered dreadfully serious for a young officer to get himself killed, due to inexperience, but for him to get experienced, sometimes irreplaceable soldiers killed unnecessarily was unforgivable. Thus one would find newly joined Lieutenants from England driving cars, acting as gunners and wireless operators, even cooking (often a cause of complaint!) before they were allowed to take over the jobs first of a Troop corporal, then of a Troop sergeant, and finally to have their own Troop. It was a system that worked wonders.

On January 21, 1942, Rommel suddenly went over to the offensive. He wrote 'After carefully weighing the pros and cons, I've decided to take the risk.' It turned out to be well worth it. The 11th were soon in the thick of the fight, and very nearly lost half of 'B' Squadron cut off when German tanks got round behind the Guards Brigade and captured Agedabya. A week later Rommel was in Benghazi, and thereafter the 8th Army was in retreat again. The original attack had found the 11th at Msus—'the Msus Stakes' became a famous desert cynicism—and until the end of the month the 11th operated between there and Antelat in a covering screen, being down to four effective Troops between the two squadrons by the end of January.

By February 8 the whole regiment had returned to the Delta to refit and rest. The last days of this long tour had been severe ones. One of the oddest incidents had concerned Lieutenant J. R. Ballingal's 5 Troop 'A' Squadron which had endured a fierce clash with Rommel's armour and anti-tank guns near Antelat before all his cars were disabled. For an hour and a half before actually captured Ballingal had sent back a stream of cool and valuable information about the enemy; he even reported himself 'certain to be put in the bag' because of the large number of tanks around him, which were being engaged by the Bays. Ballingal and his troop were put on the backs of Mark IV tanks for want of better transport and took part willy-nilly in the next German advance before being transferred that night to two lorries thirty miles east of Msus. These were full of British and Indian prisoners. One lorry broke down and had to be towed, and the German driver and guards lost their way during the night and got thoroughly confused. By daybreak it was clear to Ballingal that the Germans were not only completely lost but had been motoring in the wrong direction, east instead of west.

> 'Eventually the thoroughly frightened Germans asked my advice, which was simply that we should continue as we had begun, towards the west (I was sure by then that we were heading N.E. but that did not occur to our warders). With daylight the sky grew completely

grey and there was no sun, thank God, to show the Germans how mistaken they were.

'At about 8 a.m. I saw two or three vehicles a mile or more to our right. They were camouflaged and moving towards us but the Germans did not see them . . . we cautiously removed our berets and began to wave them over the back of the lorry. . . .' [*The Eleventh at War.*]

The cars were South African Marmon-Harringtons of the regiment's old friends and colleagues. Thus Ballingal, Sergeant A. H. Christmas, M.M., and the Troop escaped and brought other British prisoners back with them. John Ballingal, M.C., became one of the outstanding officers in the 11th for the rest of the war and after it.

Chapter 14

The End in the Desert

'Say not the struggle naught availeth,
The labour and the wounds are vain,
The enemy faints not, nor faileth,
And as things have been they remain.'

(A. H. Clough.)

THE 7th Armoured Division did not return to the desert until the beginning of April 1942, at which time the British had stabilised a front on the Gazala line. But the regiment did not go with its parent formation, for towards the end of April Colonel Leetham, on Auchinleck's instructions, took it to Iraq. The C.-in-C. Middle East was justifiably worried about the possibility of a German attack through northern Persia if things continued to go badly for the Russians, and a future link-up between von Manstein from the north and Rommel from the west would be utterly disastrous for British power in the whole Middle East.

The regiment drove over 1,000 miles to Mosul in seven days' motoring, along the traditional route Damascus–Palmyra–the Euphrates. When it arrived on May 6 Mosul was boilingly hot. After a month spent at Mosul in very different country from that to which they had become used, the 11th, as part of 252nd Indian Armoured Group of 10th Army, crossed the Tigris and moved via Kirkuk to Persia, reaching Kermanshah on June 9; high up in the hills, this pleasant town was on the aid-to-Russia supply route, but absolutely no reconnaissances were allowed by the Russians in the area of Persia they occupied.

Meanwhile, the news from North Africa was very bad. 'Now the 11th Hussars have left we can afford to take more risks,' a sentence in one captured German document stated. Whether or not this was strictly true, a mystique had grown up that if the 11th Hussars were taken away from the desert things invariably went wrong, no matter how much 'Groppi's Hussars' and 'the Gaberdene Swine' might deny it in Cairo. On June 10 the Free French, who had long courageously defended Bir Hachiem, were ordered to pull out; on June 13 Rommel won the harsh tank battle of Knightsbridge, and on the 17th reached Sidi Rezegh again. Worse still, Tobruk fell almost immediately afterwards. The soldiers of the 11th felt restless and futile so far away.

But worse was to follow. On June 23 Rommel's troops crossed into Egypt, and actually reached Mersa Matruh on June 28. Next day the 8th Army was rocking on its heels at El Alamein, final defensive position before Alexandria. Things had never been so bad. Even before all this happened an idle officer had been heard murmuring:

> *'In Persia*
> *I feel inertia—*
> *I would far, O far O*
> *Rather be in Cairo.'*

This was true of the men too, for when at last on the night of June 30 orders came to return to Cairo with utmost speed, one of them wrote, 'most men welcomed the move, for on the whole Egypt was their home and Cairo their London.'

Rommel's successes, great though they had been, had been rendered the easier by the outbreak of war with Japan. Not only were two British infantry divisions, four light bomber squadrons, and many ack-ack and anti-tank guns and other equipment which had been earmarked for the

Middle East diverted to the Far East, but from the Middle East itself ten bomber and fighter squadrons of the R.A.F., 6th and 7th Australian Divisions, and 7th Armoured Brigade were sent eastwards. The Royal Navy in the Mediterranean, by unhappy coincidence, received some crippling blows at this time, thereby rendering Rommel's supply problems less difficult, and the Luftwaffe and Regia Aeronautica renewed their attacks on Malta.

The strength of the regiment was now thirty officers, 471 other ranks, with thirty-two armoured cars and about 100 'B' vehicles. Cairo on July 7 seemed a city with a dead hand over it, with shuttered shops, wild rumours circulating among the Egyptians and even some of the British, and field guns at the ready on the Gezireh Club polo grounds. Early on July 16 the 11th set out on the familiar road to the front—a sadly short journey on this occasion—for the fifth and as it turned out last time since the war had started. They now came under Brigadier W. G. Carr's 4th Light Armoured Brigade, the King's Dragoon Guards having taken their place as divisional troops. There were now more than 200 armoured cars in 8th Army, and 12th Lancers were in the same brigade.

Between now and the battle of El Alamein the regiment, though usually in the front line, had the odd experience of being outside the main battle; during Rommel's last great effort to break through to Cairo between August 31–September 7 it was playing outside left to the Eighth Army and seeing little of the game. Enemy air attacks, however, inflicted some casualties on vehicles and men and taught everyone to be air conscious again.

Great changes had taken place both at the highest levels and, more humbly, within the regiment. Churchill replaced Auchinleck in August by General Sir Harold Alexander; Lieut-General Gott, of XIII Corps, had been appointed 8th Army Commander when to the consternation and grief of all he was killed in an air crash. He had already taken as

Imperial War Museum
Mr Winston Churchill accompanied by Lieut-General Montgomery inspects the 11th Hussars during his visit to North Africa.

his A.D.C. John Poston, who had won two M.C.s with the regiment, and his successor, Lieut-General B. L. Montgomery, asked Poston to serve him too. Poston agreed and was to stay with 'Monty' as one of his most trusted liaison officers to the end—or almost, for in the last week of the war he was killed by a rifle bullet in Germany, driving through a forest which still contained German troops.

On September 7 'C' Squadron formed the first Jeep troop in the regiment, and at the end of the month an ack-ack troop came into existence. It consisted of one Breda gun, one German 20 mm. and one Flak truck with twin machine-guns.

Improved Humbers were issued to replace those which were fading away after the long Iraqui-Persian expedition, and after Alam Halfa until Alamein patrolling alternated with training. From September 18 Major-General A. F. Harding took over 7th Armoured Division. The little scarlet Jerboas or desert rats painted in a white circle on all the cars, and worn on shoulder flashes, had still a long way to go, but the lavish influx of new weapons and equipment that continued through October, including 300 most welcome Sherman tanks from America, amazed those who had grown used to fighting their war on a shoestring. Morale, which had sagged badly in 8th Army after its defeats and 'push-me-pull-you' locomotions of the past year, improved enormously as a result; but also due to the superb self-confidence of Montgomery, who, however flamboyant his methods, undoubtedly injected confidence into others and rekindled enthusiasm even among the most cynical.

The story of El Alamein and the subsequent pursuit to Tunis has been told many times, and here only the last stages are recounted from one regiment's point of view. It should merely be said that the traffic control arrangements in the south through the enemy minefields on October 23–24 were in the capable hands of Major Lawson and 'A' Squadron, helped by two troops of the 2nd Derbyshire Yeomanry.

At last, on the night of November 3 and the morning of the next day, the breakthrough came for 7th Armoured Division, led by the 11th and 22nd Armoured Brigade (Brigadier G. P. B. Roberts). Now was the chance for real movement, after they had been hemmed in for ten days by innumerable minefields and thousands of vehicles, with artillery behind them sometimes hubcap to hubcap firing thousands of rounds in the biggest barrage heard in Africa. There were to be many tough actions and many delays, not all of which seemed justifiable. But the advance ground on. Sometimes rain and mud slowed it down and even bogged down cars and tanks of both sides. As always, unexpected things happened. 2 Troop 'B' Squadron ran straight into an enemy column retreating west as fast as it could go, on November 5, and knocked out two eight-wheelers, one 88 mm. anti-tank gun, and one 105 mm. gun, with the help of 3 Troop. Next day another column of the 90th Light ran slap into the rear of 22nd Armoured Brigade, went right through it nearly disposing of 2 Troop 'B' Squadron in the process, and disappeared northwards almost before anyone knew what was happening. The poet Keith Douglas, then commanding a Crusader troop of the Sherwood Rangers—who referred to the regiment as 'Our red friends on ponies, the Cherry Ps'—related (in his posthumous book *Alamein to Zem Zem*) an incident near Mersa Matruh:

> 'We came up to the line of high ground without further excitement, though we put a six-pounder shell across the bows of one of the 11th Hussars' armoured cars, which had moved out well ahead of us, and was under suspicion of being an enemy vehicle. We aimed well ahead of it, and the effect was immediate. It came tearing towards us like a scalded cat and drew up alongside. Its commander, a sergeant, leaned out and said "Was that you firing at us?" We admitted it, and apologised. "With that?" pointing to the long snout of our six-pounder. "Well, yes," we admitted. "Phew!" he said. "Don't do it again, please." He moved out ahead of us again.'

On November 12 'B' Squadron sighted the Tobruk defences, and next day Second-Lieutenant R. G. G. Copeland's troop was first into the town. The tanks of 22nd Armoured Brigade were unable to keep up with the fast retreating enemy, and were always an hour behind, for they had completed their viable mileage long before the offensive started. On November 15 the 11th gazed at derelict vehicles as far as the eye could see on the Knightsbridge battlefield of the previous June; two days later 'B' Squadron reported Msus clear. On November 20 'A' Squadron were first into Benghazi.

By December 16 'A' Squadron had reached Marble Arch, and from the 19th the whole regiment had a week's rest by the sea in Tripolitania. For a considerable part of the previous six weeks Colonel Smail had been commanding a column consisting of the 11th Hussars, one Battery 4 Field Regiment R.A., B Company 1st Rifle Brigade, a troop each of Royal Engineers and anti-tank guns, two troops of ack-ack guns, and an R.A.S.C. Company.

During their lull, Colonel Smail and Major Lawson dined at 8th Army headquarters with General Montgomery, and future armoured car policy was discussed. The General opined that armoured cars should not be used for fighting, but only to get information, a statement which caused some wry amusement however sound in theory.

On January 15, 1943, the advance to Tripoli began; soon the going, with not only mud but wide, deep treacherous wadis with very steep sides, got worse than anything met before. But by January 22 the enemy rearguard had withdrawn to Castel Benito, and during the night were cleared out by 131 (Queens) Brigade. At 3 a.m., well before first light on January 23, Sergeant Lyon of 'B' Squadron, followed by Captain R. R. Lockett, led the way through Castel Benito to Tripoli itself, and after a cautious approach found the town to be undefended and entered it at 5 a.m., well before the tanks of 23 Armoured Brigade with 51st Highland Division, which entered from the east. The docks and harbour were wrecked by bomb damage and demolitions, and it was impossible to get close to them as the approach road was blown. At first Tripoli seemed a dead town whose inhabitants struck the 11th Hussars as 'just like gnomes coming out of underground dwellings'. But once they realised that the armoured cars were British they went wild with enthusiasm, and the regiment had its first taste of 'Liberation'; festivities only increased when the skirl of Highland bagpipes was heard, coming closer and closer to the Piazza d'Italia.

On January 29 Lieut-General Sir Oliver Leese, now commanding XXX Corps, sent this message to Colonel Smail:

> 'Congratulations on winning race to Tripoli. A fitting climax to all your excellent work.'

But more was still to come.

Colonel Carver, writing in 1943, said that 'The immediate result of the capture of Tripoli was an outbreak of Tripoli fever; in other words more attention was paid to Tripoli than to the enemy.' 7th Armoured Division recovered from it about midday on the day of Tripoli's capture, he thought, but 'Others continued to suffer from it for several weeks.'

Rommel, who considered that 'Montgomery had an absolute mania for always bringing up adequate reserves behind his back and risking as little as possible', thought that his opponent showed 'real stature' at Tripoli.

The regiment pushed on through Bianci westwards, and there found a placard by the road which read 'Goodbye and keep smiling—Ramcke.' This veiled threat came from Major-General 'Papa' Ramcke—Eisenhower was later to describe him as 'a formidable figure'—who commanded an élite brigade of Luftwaffe parachutists.

On January 25 the regiment handed over to 12th Lancers and settled down at Bosco Mussolini for a month completely to refit. The Humbers were worn out and new Daimler armoured cars, with a 2-pounder and 7·92 mm. Besa coaxially mounted were ferried from Benghazi. With these fine cars, the best they had ever had, the 11th were to go on to war's end. Equally valuable was the new Daimler 'Dingo' scout car for two men (the Daimler armoured car carried three), a versatile little car which was most manoeuvrable. As the driver's seat was slanted slightly sideways it could be driven backwards at speed; so too could the Daimler, with more hazard and less quickly, for the commander had to contort himself crouching down, and steer, looking through a slit, with the reverse steering wheel, shouting orders to accelerate or slow down to his driver, although he had a hand throttle. Constant practice at this startling manoeuvre, during which the commander was lucky if he did not get hopelessly entangled in wireless flexes and the general paraphernalia of the turret, paid off handsomely in Normandy and beyond.

The Dingo carried a mounted Bren gun, later replaced by twin Lysander K-Guns from the R.A.F., marvellous guns with a very high rate of fire, the Bren then being carried loose as an extra weapon. American White Scout cars were also issued, proving invaluable for carrying 'White sections' or Troops, which the 11th Hussars used for dismounted infantry work, usually with a sabre Troop, sometimes on their own. Finally more Jeeps were issued. Thus each squadron was now reorganised to consist of four armoured car troops each with two Daimlers and a Dingo, the White Scout Troop with four Whites and a Dingo, and a Jeep troop with three Jeeps. With squadron headquarters each had a strength of about 120 officers and men.

The regiment, in fact, was reorganised for a new kind of warfare in much closer country than before, the woods, hills, tracks and rolling farmlands of Tunisia, very different from the desert.

The Prime Minister arrived in Tripoli on February 3, and with him the C.I.G.S., General Sir Alan Brooke, the C.-in-C. Middle East, and General Montgomery. Mr Churchill, who was deeply moved, reviewed the troops of the 8th Army that day and the next. He told Lieutenant G. J. Lovett how proud and glad he was to have a personal escort troop of 11th Hussars.

On March 7, 1943, the regiment rested, re-equipped, and having acquired a taste for the local Chianti, left Bosco Mussolini, just in time for the final advance, it thought, though a fortnight was to elapse before General Montgomery lifted the starting gate. Next day it was back under 7th Armoured Division three miles north-west of Medenine, where heavy fighting had been going on, and Rommel's last big effort to break the 8th Army's front (which he himself admitted came a week too late) had been beaten off with severe losses on both sides.

Also on March 8 Major Lawson, who had commanded 'A' Squadron for two years, was ordered specially by Montgomery to leave the regiment and join 1st Army so that he could instruct the Americans in the best use of armoured cars and armour. The Army Commander told Colonel Smail that he wanted to send General Eisenhower 'the best squadron leader in the 8th Army'.

Captain Reid-Scott, just returned from Wavell's staff in India, took over. The night before he had had the privilege, with Colonel Smail of dining with Montgomery at his field headquarters: he recorded that 'both were disappointed in the low level of conversation, which was chiefly about Army wives—a subject on which much better information is always available in the 11th Hussars Officers' Mess.'

The division was now commanded by Major-General G. W. E. J. Erskine, who had taken over from General 'John' Harding, who had been wounded just before Tripoli when standing on Lieutenant R. E. Wingfield-Digby's Humber observing the battle.

Rommel, who was physically sick and mentally exhausted, left for Rome and Berlin on March 9; he had been ill for some time, but even now he could inspire the Afrika Korps. His A.D.C. wrote to Frau Rommel at the end of February:

> 'It was wonderful to see the joy of his troops during the last few days, as he drove along their columns. And when, in the middle of the attack, he appeared among a new division . . . right up with the leading infantry scouts in front of the tank spearheads, and lay in the mud among the men under artillery fire in his old way, how their eyes lit up.'

Rommel handed over command of Army Group Africa

to General von Arnim. There were by now a number of Panther and Tiger heavy tanks in the Panzer divisions, armed with high-velocity long-barrelled 75 mm. or 88 mm. guns. The one drawback to these, as was later found in Europe, was that they were slow and not very manoeuvrable. One tended not to appreciate these deficiencies when coming up against one in battle.

Near the Mareth Line 'C' Squadron, now under Captain W. V. Burdon, M.C., lost four cars blown up on mines and heavy shellfire added to its casualties in officers and men.

On April 8 'C' Squadron, between Mahares and Gafsa, got some of its own back by capturing the first Tiger (Mark VI) that the division had encountered. It had been damaged, apparently, by shots from 8th Armoured Brigade and left behind as a pill-box. 2 Troop, not knowing this, attacked it from 500 yards range and its crew baled out and surrendered the Tiger. The same day 1 Troop captured 100 Italians and three 105 mm. guns.

On April 9 'A' Squadron headquarters distinguished itself by over-running an abandoned Italian workshop containing forty-two light tanks. The enemy now was pouring west, and next day, April 10, the squadron entered Sfax at 0750 a.m. Exactly an hour later a triumphant Highland Division officer shouted to Captain Crankshaw, as squadron headquarters entered the town, 'Well, if the 11th Hussars were first into Tripoli, at least they didn't beat us for Sfax!' Crankshaw informed him that one of 'A' Squadron's troops was already ten miles up the road.

On April 30 7th Armoured Division was ordered under command of 1st Army and moved via Kairouan to Le Krib. In Tunisia now, waiting for the final advance, the cars' camouflage had been changed from the yellow and pale green desert style to slightly darker colours. During all this time the British had more or less complete air superiority. Captain Reid-Scott had registered his surprise at the time of the static Mareth Line patrols on waking up one morning to find about seventy aeroplanes on the aerodrome just west of Medinine. For apart from a little night bombing no German planes had been seen in the area for days. Near Gabes on April 3 the R.A.F. shot down fourteen Stukas out of twenty, and the ack-ack brought down a night bomber in flames; and on April 15, when the regiment made an approach march along very congested roads, no one seemed to bother about air precautions, which only showed how little the 11th had been bothered lately.

At Le Krib the division had come under command of IX Corps, just taken over by Lieut-General B. G. Horrocks. This Corps breached the main German position near Medjez el Bab on May 5, with V Corps on its left taking important high ground which could have dominated the division's advance if not captured. At 3 a.m. on May 6, a tremendous artillery barrage was laid down in front of the two infantry divisions, and the R.A.F. made raid after raid against the German positions in the Tunis area with light and fighter-bombers. Both infantry divisions gained their several objectives and then 22nd Armoured Brigade advanced through 4th Indian Division and 26th Armoured Brigade on their right through 4th Division. After some opposition 22nd Armoured Brigade reached Massicault and the Djebel Achour, the high ground to the north, and patrols pushed on to the outskirts of St Cyprien. 131 Brigade followed on, securing the ground taken by the armour.

At first light on May 7 the advance was resumed, and numbers of German tanks and 88 mm. anti-tank guns in the St Cyprien area and on the ridge overlooking Tunis were dealt with. Now it was time for the armour to dominate all roads leading into Tunis from the west and for the 11th to carry out close reconnaissance, although the division's right flank was very open as 26th Armoured Brigade was some way behind. There were groups of enemy tanks on this flank but 1st R.T.R. got behind them and successfully engaged them. Soon after 3 p.m., the time at which 22nd Armoured Brigade first looked down at Tunis itself, General Erskine ordered it to go into the city.

The 1/7th Queens were rushed up to join the armour because 11th patrols had reported that street fighting might be expected. Nobody really knew what to expect of Tunis.

1st R.T.R. began the advance into the city straight down the main road with 5th R.T.R. on their left and 4th C.L.Y. on their right. Just east of Manouba the 11th Hussars took over with 'B' Squadron in the lead and by 3.45 p.m. Lieutenant Burridge's Troop was right inside the city. The Signals log of 22nd Armoured Brigade records:

'"B" Squadron 11th Hussars report right patrol now seems to be right in town itself. Can 1st Royal Tanks send something forward to help? It is raining steadily, the Troop is surrounded by surprised Germans firing at the cars, hundreds of others surrendering, wildly excited civilians blocking way, showering flowers and pressing with wine and other offerings.' [*Quoted by Verney.*]

Colonel Carver duly sent part of his light squadron to help rebuff the Germans and control the French, who were going wild with joy. All troops of 'B' Squadron had to fight their way into Tunis, and when 'C' arrived to help they too had much street fighting, during which hundreds of prisoners were taken and Lance-Sergeant R. T. Nash engaged and knocked out a 75 mm. anti-tank gun. The Germans had been taken completely by surprise in almost every case, many of them were walking about the streets with their girl friends, one officer even rushed out of a barber's with his face still covered with lather. By dusk the civilians with captured arms were almost as dangerous as the Germans, but the Queen's Brigade mopped up most efficiently and by morning of May 8 the last remaining pockets of Germans had surrendered, to swell the thousands of prisoners already taken.

'A' Squadron, meanwhile, had had rather a depressing day, guarding the divisional left flank in the Djedeida area, where there were still large and aggressive numbers of Germans about, well armed. Having lost one Daimler blown up by a double Teller mine they then lost two Whites and Lieutenant J. Garrard's Daimler, set on fire with its 2-pounder blown off, to four well concealed 88 mms near the railway four miles east of Djedeida. Two Troops and the White Troop with two mortar sections had nearly run straight into these guns, which were protected by 150 infantry in trenches. Lieutenant D. A. F. Williamson and Sergeant J. A. Hall had been conferring on a narrow road where it was impossible to turn round when they actually heard fire orders being given in German from a clump of trees less than 400 yards away. Miraculously most of the cars managed to get off the road by driving straight ahead or reversing, but Garrard's crew and eight of the Whites had to take refuge in an adjoining cornfield, where their every move produced raking fire from machine-guns. Garrard was reported killed, but after pretending to be dead reached a safe hiding place by wriggling through the corn and turned up safely next morning. Sergeant Fitzpatrick also got away with most of his Whites section by crawling away in the dark while the Germans were searching the cornfield. Next morning the C.L.Y. captured this German position after losing two Shermans, and recovered three 11th Hussar prisoners. 'A' Squadron then restored its morale by taking 330 prisoners, many from the Jaeger Regiment of the Hermann Goering Division, and made contact with patrols from the U.S. 2nd Corps, which had taken Bizerta. At the Medjerda River, too wide to cross, twenty-five members of the 15th Panzer Division on the far side, under a very fat *Oberleutnant*, were persuaded by a burst of fire to surrender, and had to pay Tunisian peasants 50 francs a head to cross over on farm horses. The *Oberleutnant* proved overweight for his mount and turned turtle in mid-stream.

'C' Squadron on May 8, also on their way to this river, captured an Italian searchlight battery, the Town Major of

Tunis and his staff, and twenty lorries overflowing with Italians. 4 Troop alone took 1,000 prisoners during the day, and 1 Troop a German aerodrome, where it assimilated three mobile 'Chicago pianos' (four coaxially mounted German 20 mms on trailers, which were very effective ack-ack weapons). North and east of Tunis the enemy were hopelessly trapped waiting for sea transport which never came, and 'B' Squadron put up the record by capturing 10,000 Germans and Italians.

Next day, May 9, was the last day of active operations for the regiment in North Africa. 'C' Squadron had been ordered to cross the Medjerda, but all the bridges were down and fording proved unfeasible. Hundreds of Germans were waiting on the far bank to surrender, so Van Burdon organised a horse and mule ferry with the help of local Arabs—still charging 50 francs a head—after crossing in a native mule cart. During this crossing he fell out with the owner-driver and dumped him in the river 'for refusing to obey orders and driving badly anyway', to the amusement of British and Germans alike. Some 200 rather damp Germans were ferried across and sent back to Tunis, including the former Chief Ordnance Officer of Tripoli, a rather drunk but definitely pro-British Austrian, who owned a château whose hospitality after the war he offered.

'A' Squadron liaised with the Americans at Mateur and 1st Division near Djederda, and managed to find a crossing over the Medjerda at last, becoming the only troops in 1st Army to operate across the river. Captain G. V. Churton, who had joined the regiment three months earlier from the Cheshire Yeomanry, led the Americans near Protville north up the main road to Bizerta. The German radio had supposedly announced that resistance in North Africa had ended the previous night, and large numbers of Germans were waiting to surrender in the Porto Farina area. Many others, however, were busy destroying equipment and blowing up installations in the town, some three miles away, whence enormous explosions could be heard. It was imperative to get there as soon as possible to stop this destruction, but the Americans behind Churton and Sergeant Hall suddenly halted. When asked the reason they told Churton that an attack on a pre-arranged plan would have to be put in with an artillery barrage, followed by an infantry attack with two battalions. As the American infantry were still miles away, this seemed impracticable as well as unnecessary. The American armoured commanders agreed rather unwillingly to follow Churton and Hall into Porto Farina, where resistance was expected since by no means all the Germans had heard they were supposed to surrender and many were ignoring the rumour. By 10 a.m. both troops were inside the town, at which moment the Americans put down several rounds of gunfire into it! Inside Porto Farina there was a huge traffic jam, and the armoured cars followed by Shermans of the American Armoured Division (with which Lieut-Colonel Lawson was travelling) had a hard job negotiating the narrow streets with a mass of Germans trying to come the other way and hundreds of civilians surging round, singing and cheering themselves hoarse.

There was a great deal of German equipment, most of it by now burning furiously, including tanks, guns and transport. This was interesting as a British Intelligence summary a week before the final attack had stated that the Germans had only twenty-four 88 mm. guns left in the whole of North Africa. Captain Reid-Scott had counted forty along the road between Chaouart and Porto Farina alone.

An hour later, when Churton found he could not get his cars on to the beach north of the town, he walked along it, armed only with a revolver, accompanied by a *Hauptmann* of 15th Panzer Division and an interpreter. The beach was teeming with Germans, many of whom were rather pathetically making rafts. During the next three hours Churton, single-handed, collected about 9,000 prisoners, whom with some pride he organised and handed over to the Americans.

This was the end in Africa. The regiment had gained the so-called triple crown, first into Tobruk, Benghazi and Tripoli, indeed a quadruple one if Tunis was counted. It had only just beaten Payne-Gallwey's 1st Derbyshire Yeomanry of 6th Armoured Division into Tunis, by such a short head that it was glad to share the honour.

It was really an honour for the 7th Armoured Division, and considered as such. The Desert Rats had come 2,000 miles in 180 days, hardly ever out of contact with the enemy.

The effect of the North African victory, after depressing years of seemingly fruitless struggle, was electric in Britain. Even the staid leader columns of *The Times* were moved in the general rejoicing to refer to 'those incomparable paladins, the 11th Hussars'.

Tributes such as these were very welcome, but perhaps more appreciated was a German Intelligence report that had been captured in September 1942. An *Oberst-Leutnant* G.S.1 (A) referring to a German reconnaissance unit which had allowed its office lorry, containing operational orders, to be captured, had written:

> 'It is to be regarded as a decided exception due to the inexperience of this unit; it would never have occurred with the 11th Hussars.'

And most of all would they have liked Rommel's later tributes to the 7th Armoured Division and to themselves. The Field-Marshal was often critical not only of British senior commanders but of the aggressive capabilities of some British units; he also thought many British troops inadequately trained and too inflexible. 'An exception was to be found in the British reconnaissance units, whose training was first class,' wrote Rommel, and he singled out for honourable mention generally the 7th Armoured Division, particularly the two rifle battalions of the Support Group (1st K.R.R.C. and 2nd R.B.), the 11th Hussars, and the artillery (R.H.A.).

Chapter 15
Defeat in the West for the Germans

'The Desert Rats will shortly become water rats. . . .'
(Lord Haw-Haw broadcast, picked up in Italy, December 1943.)

KING GEORGE VI visited North Africa in June 1943, and although he had a very tight schedule and had to cover great distances, he expressed particularly a wish to see the 11th Hussars rather more informally than many of the other regiments which he had to inspect. He was, after all, Colonel-in-Chief of the regiment. It was then at Homs in the Tripoli area, the regimental home for more than three months before the invasion of Italy. King George, after reviewing the 7th Armoured Division at Olivetti, twenty miles west of Tripoli, came to his own regiment the next day. With so many Allied troops to see during his short time in Africa this was a gesture much appreciated by the regiment. King George had had a long and tiring day, but he must have been moved to hear the tremendous cheering after he had inspected the regiment, not in an open car but on foot.

Often he stopped to talk to troopers and N.C.O.s, and before he left he reminded his soldiers that it was eleven years since he had last inspected them at Tidworth in 1932.

On September 4 the regiment began its move to the Tripoli assembly area for the coming invasion of Italy, already wholesomely foreshadowed by success in Sicily. Waterproofing was carried out, and, on September 9, eight French 75 mm. guns mounted on American White half-tracks were issued to the regiment. These were to prove invaluable in close country work, used as close support artillery, one Troop to a squadron. They were much used in Italy, and far more so in Normandy, Belgium and Holland, although in 1944 they were different in the sense that the 11th had handed over its original 'Gun Troops' to the Canadians in Italy and had been issued with replacements courtesy of the Americans. An armoured car regiment did not, in those days, have much heavy fire power, and the effect on the Germans when a single squadron brought into play two very accurate French 75s 'up at the sharp end' was, if nothing else, confusing. As these guns could, and sometimes did, fire over open sights, they not only persuaded

A gun troop firing their French 75 mms mounted on American White half-tracks in Italy.

the enemy that much heavier opposition was coming up against them, but also made them wonder if they had mistaken one or two probing patrols for the advance guard of a powerful force of tanks and artillery. This was excusable, since the gun troops could fire at ranges varying from 200 to 8,000 yards, and many times their French designer was silently praised.

'C' Squadron embarked for Italy on September 16, and three days later the rest of the regiment sailed to Salerno. Until October 2 'C' Squadron worked with 131 Brigade, and then it went into regimental reserve. The 11th were full of confidence after their long rest, and when they left Africa for Italy it was generally felt that they had 'never set out on an operation better prepared'. Their stay in Italy, however, was to be brief. What it taught them, and it was a most rewarding lesson, was that armoured car work in Europe was totally different from that in the Middle East. If Italy did nothing else, it prepared and trained the regiment for the future close-country warfare in Normandy, Belgium, Holland, and Germany.

On October 6, for example, the regiment was operating in enclosed country: thick vineyards, visibility impossible over 1,000 yards even from house tops. It was good to be back in Europe, but European weather had been forgotten after nine years in Egypt, Palestine and the desert. Rain, mud, hot sunshine followed by very cold nights, mist, fog, and mountains, above all, were a different cycle from the burning heat of the daytime desert followed by the chill nights. Not surprisingly, many men went down with malaria, sandfly fever, dysentery, and even jaundice.

The Luftwaffe again made not few but many appearances, for with shorter lines of communication Italy was still very much their province. Italy had accepted terms of surrender to the Allies on September 9, but Field Marshal 'Smiling Albert' Kesselring, himself a Luftwaffe officer, was not likely to let that perturb him. Indeed, his defence of a hostile Italy until close to the end of the war was a great military accomplishment.

By September 28 the two infantry divisions of X Corps had cleared the Cava di Tirrani pass on the way to Naples, and the 11th Hussars followed slowly up the centre line through Nocera and Angri almost to Scafati. There it was decided to pass 7th Armoured Division to the north of Mount Vesuvius instead of along the coast road to Naples. Nobody could have been more pleased than the 11th when it was heard that the King's Dragoon Guards had been first into Naples.

West of Vesuvius on a line facing north, the regiment deployed with 22nd Armoured Brigade to advance to the Volturno River. The plain before the river was wide, a memorial to this day of Mussolini's *Opera Nazionale Cooperativa*, open farming country studded with *Onc-Houses* and small holdings, where one met bullocks drawing farm carts and listened every evening to the Italians praying to the Virgin Mary for good fortune to the Inglesi and the reverse to the Tedeschi.

On October 3 a force from 7th Armoured Division, including 'B' Squadron, 11th Hussars, was all grouped in the main square of Arzano, where the British soldiers got a magnificent reception from the Italians. But in nearby Giuliano, where Italian patriots had shot two German soldiers, the Germans had retaliated by shooting ten Italians—not unjustifiably. But while these Italians were being buried, not long after the 11th armoured cars had passed through the village, the Germans mowed down with rifle and machine-gun fire the whole Italian crowd round the cemetery. The war was getting nastier. There had been two incidents in the desert which the 11th had not forgotten. One was when some twenty Italians had put up their hands to a Troop and had then used grenades and bullets against the 11th as they went forward to accept surrender. Those fools were mostly killed; another was when a German sergeant, whose officer had surrendered with a group, tried to knife the Troop sergeant. He was immediately shot, but, curiously, 'refused to die'.

There were other frustrations in Italy for the regiment, whose tradition it had always been not to make their leading car a map reference. It is recorded that on October 24 a British infantry division appeared slow and 'not pushers'. For the past fortnight their patrols had never been within two miles of their map references. The 11th might be new to European fighting, but they knew their maps and how to fight. Having so often been shot at by 'own troops' in the desert—an occupational hazard—they did not take that sort of thing amiss. But they liked to have efficient friends behind them, upon whom they could depend.

Mines were becoming an increasing problem in Italy, and after several disasters it was rightly decided to sandbag the inside floor of Dingos, usually the leading car in close country. This move saved a lot of lives in the next eighteen months.

At one time 'A' Squadron had a troop of self-propelled 'Priests' (105 mm. guns) attached to it, on loan from the Royal Horse Artillery. These were exceptionally useful; but by October 5 the rains had begun in earnest, and the advance was delayed all along the front. 7th Armoured Division was under command of the United States 5th Army, and was divorced from Montgomery's 8th Army on the eastern side of Italy. The division as a whole got on splendidly with the Americans, who had learned a lot since the days of Kasserine and Tunisia. The 11th learned soon that the Americans of 'General Mark W. Clark's 5th Army, somewhere in Italy', were excellent fighters, and not just a convenient by-line for American press correspondents.

The 11th did a lot of patrolling but not much heavy fighting in Italy. Some genius, either a General or a member of the War Cabinet, rightly decided that they and 7th

Armoured Division should be exposed to European warfare in Italy before they went to Normandy. This was correct. It is not true to say that the 11th did not do much valuable work in Italy, but it is truer to say that Italy opened the regiment's eyes to the totally new hazards of close country warfare. The frustrations met by squadron leaders used to open warfare in Africa became evident in Italy. New officers and soldiers who had been trained in England took this sort of thing in their stride, and perhaps they contributed a healthy percentage of what they had learned without the benefit of fighting. But they gained much more from the experience of the regiment.

The Germans mostly relied upon heavy shelling, a few anti-tank guns and self-propelled artillery, and the occasional tank, to supplement their extremely good infantry. The regiment did its best with patrols, and had some hard actions.

The Italian locals were very helpful everywhere, often risking their lives in order to run and warn a patrol that there was a German anti-tank gun or something round the corner; these civilians were often much braver than the Italian Fascist Army of Mussolini's boasted 'One million bayonets' which the 11th had met in North Africa.

Grazzanise, the Volturno, Cancello, the approaches to the Roca di Mondragone and the Garigliano are well-remembered names. But it seemed to be an infantry campaign, and it was not without relief that the regiment heard, first from the local Italians, of course, that it was going home to prepare for the invasion of France.

During the voyage, Officer Commanding Troops on the *Cameronia* succeeded in doing something not otherwise achieved except by German artillery and Luftwaffe bombers during the fighting round Caen, in Normandy. He made 11th Hussars wear tin hats—during boat drill.

Dolce fare niente did not suit the 11th Hussars in Italy.

In France, and onwards to Hamburg, despite frustrating gaps, there was to be a great deal more and fiercer action.

★   ★   ★   ★   ★   ★

In England the regiment refitted again, having left its cars and gun-troops with the Royal Manitoba Dragoons in Italy. New White half-tracks with French 75 mms were issued, and although the log books showed that some of these guns had barrels dating back to 1910–12 they had evidently been refurbished. At least these guns, two to a squadron, and occasionally batteried, did sterling service to the sabre Troops, and it was only after frequent 'cook-offs' and an obvious loss of accuracy that they were written off just before the 11th entered Germany.

Command in Normandy, and thenceforth to the end of the war, was entrusted to Lieut-Colonel Bill Wainman. On February 24, 1944, the Supreme Commander, Allied Forces Europe, General Dwight Eisenhower, accompanied by his deputy Commander-in-Chief, Air Marshal Tedder, and by General Sir Bernard Montgomery, visited the regiment, now up to its neck in putting 'Bostic' on its vehicles in case of a wet landing. When the second-in-command welcomed Eisenhower, the American replied, with the simplicity that endeared him to most: 'Good morning. My name's Eisenhower.' As if one didn't know. But Eisenhower had a very professional way of inspecting troops, and after asking several personal and pertinent questions he seemed to be satisfied. Anyway, he took a glass of port and a slice of plum cake in the Mess. Montgomery, also obviously pleased, enjoyed the cake too. Years later Eisenhower was to write to an American friend:

> 'I well remember the day I inspected the 11th Hussars; that regiment had a proud record and I considered it a great personal honor to have a visit with them before the Invasion.'

The interesting thing about that letter is that it was dictated—and signed 'D.E.'—from the White House, Denver, Colorado, on November 2, 1955. At that time President Eisenhower was in an oxygen tent recovering from his first heart attack. The White House, for Presidential reasons, is where the President is.

★   ★   ★   ★   ★   ★

So much has been written about the fighting from Normandy to Hamburg that it is, perhaps, better to give one 11th officer's impressions of it. The dangers and the acts of courage by many people are not mentioned much, but the spirit of the regiment emerges. The 11th Hussars had at times tedious duties to perform, almost like their forbears in the Great War. But they saw far more action, and, having adapted themselves to close country work, often found a way for 7th Armoured Division, sometimes for their Corps, and not infrequently for General Dempsey's 2nd Army. They were brilliantly led by Colonel Wainman and his squadron leaders, just as they had been brilliantly led in the desert.

Although basically the reconnaissance regiment of 7th Armoured Division again, the regiment served under every British Corps, one Canadian Corps, and had frequent contact with the Americans, who, having had their battle experience in Africa and Sicily, were excellent. So many acts of gallantry and skilful judgement were displayed by 11th Troop officers, sergeants and men that it is impossible to enumerate them. The worst thing that occurred to 7th Armoured Division was the dismissal of its commander, General 'Bobby' Erskine, his C.R.A., Brigadier R. Mews, and 22nd Armoured Brigade commander, Brigadier W. R. N. Hinde. This happened after the division had failed to make a satisfactory advance in the area of Aunay-sur-Odon—after the Guards and 11th Armoured Divisions had lost about 150 tanks. 7th Armoured lost some fifty

Berlin—Humber scout cars leading Daimlers into Berlin, July 4, 1945.

itself, before General Erskine called off the attack.

When patrols of the 11th eventually went through Aunay-sur-Audon they had to call on bulldozers to remove the rubble, behind and sometimes within which German 88s had taken a fearful toll. There were exactly two buildings standing—the town hall and the biggest church.

Rommel had said in desert days that one of the biggest mistakes that the British made was continually to be changing their Commander-in-Chief. This did not happen in Normandy, but it was a grave mistake to deprive 7th Armoured of its three most experienced commanders of armour and artillery.

The fighting between Normandy and Hamburg was usually bitter and dangerous. Going down straight roads with ditches or *bocage* on either side is not always amusing, and the Germans on the whole were determined and very good marksmen—whether handling rifles, Spandaus, anti-tank guns, or Bazookas and *Panzerfausten*. But the fighting had its compensations.

One troop leader, A. D. Hunter, M.C., had a dramatic day. 'At Buxtehude an elaborate attack by the "Skins", the Rifle Brigade and a Marine Commando had been laid on, as the naval base there was thought to contain all the mine plans for the North Sea. However, Tony got there first and so "smartened up" the town's outskirts and barracks that the Commandant surrendered before the main attack started. In the bag were the Second Admiral of the North Sea Fleet, 400 German Wrens, the whole Officers' Mess cellar, and 8,000 shot-gun cartridges.' *Embarrasse de richesse*, as someone commented.

Apart from hundreds of enemy casualties inflicted and prisoners taken, thousands when every action is considered, the regiment had this tally of German equipment destroyed or captured during the 1944–45 campaign. It errs on the modest side, and omits all 'soft' vehicles.

'C' SQUADRON

S.P. gun overrun on August 5, 1944 at Aunay-sur-Audon.

Mark IV Special captured after shelling on August 18, 1944, at Reveillon.

Half-track 40-mm. knocked out on August 24, 1944, near Brionne.

105-mm. knocked out by shelling on October 17, 1944, near Kirkdriel.

Focke-Wulf 190 brought down near Lembrucht, April 5, 1945.

S.P. 75-mm. overrun on April 10, 1945, near Wildeshausen.

Armoured half-track 'brewed' on April 16, 1945, at Neuenkirchen.

'A' SQUADRON

Panther tank on August 20, 1944, at Livarot.
Armoured car captured on August 24, 1944, at Lieury.
75-mm. gun knocked out on September 1, 1944, at Airaines.
88-mm. gun knocked out on September 29, 1944, at Hees.
50-mm. gun captured on March 27, 1945, at Brunen.
Multi-20-mm. gun knocked out on March 27, 1945, at Raesfeld.
Half-track knocked out on March 29, 1945, at Borken.
Half-track knocked out on March 30, 1945, at Winterswijk.

'B' SQUADRON

Six 105-mm. guns captured intact after shelling on August 22, 1944, at St Germain de Livet.
105-mm. overturned by shelling on August 17, 1944, at Bourneville.
Two 20-mms. knocked out on August 27, 1944, at Bourneville.
88-mm. knocked out on September 6, 1944, at Nazareth.
Half-track 20-mm. 'brewed up' on September 6, 1944, at Deynze.
Messerschmitt 109 shot down on April 1, 1945, at Elte.
20-mm. gun knocked out on April 3, 1945, at Elte.

'D' SQUADRON

Two armoured half-tracks 'brewed' on June 13, 1944, at Cahagnes.
88-mm. S.P. gun captured on August 2, 1944, near Cahagnes.
Twin 20-mm. Portée 'brewed' on September 7, 1944, at Lokeren.
75-mm. A.T. gun with half-track 'brewed' on April 12, 1945, at Bassum.
80-mm. destroyed on April 18, 1945, near Soltau.
37-mm. gun destroyed on April 21, 1945, near Buxtehude.

'C' and 'D' Squadrons had landed on June 9, D plus 3, and early been thrown into the battle. The first Villers

Bocage push was both spectacular for the onlookers and startling for those taking part in it; indeed, at one time R.H.Q. Tac. were resolutely preparing to repel the *élite* of the 2nd Panzer Division with their wooden guns, while the Signals Officer, whose gun was at least made of metal, contributed his slight share to the barrage of the R.H.A., but not before spending a hectic twenty minutes getting the grease out of his barrel.

The Villers 'Box' quickly became famous, and the regiment was perhaps lucky to escape so lightly while itself writing off much German kit. Little places like Cahagnes, where a striking success was scored, and the charming but deathly village of St German d'Ectot, where the reverse unfortunately happened—these names will not soon be forgotten. The Normandy countryside with its thick *bocage* was difficult for armoured cars, and there was no quick scope for them in the slogging type of fighting that followed.

When at length the regiment did move, it crossed the Orne by 'London Bridge'. It was the worst summer for sixty years, the French said. The 11th moved backwards and forwards from Caen, and there were switches of plan, while the Americans on the right were capturing St Lo, Mortain, and Le Mans. Those who saw the wreckage of Caen and the factory area will not forget them, nor the days spent there. There was no actual operating, but much valuable liaison was done.

Besides the rain, which drenched everything, there was the dust when it did not rain, which was equally unpleasant, and turned cars and faces and berets a misty grey. And the rotting crops, the dead cows and horses, to say nothing of the dead men, the mud, the traffic jams, the abandoned or ruined kit of two armies sprawling over the fields of Normandy, all of these come vividly to mind.

Gradually the break-out got under way:

> 'Jort, St Pierre-sur-Dives, and the Poles gabbling unintelligibly on our frequencies, but fighting like lions,

then Livarot, Lisieux, Fervacques and up to the River Risle near the Forêt de Montfort. For the first time we were cracking on and not coming back, and it was exhilarating. The guns, who were firing nineteen to the dozen, found it hard to keep up, let alone the echelons. Livarot had a timely Calvados brewery, and with a quiet pull at the bottle we rushed off to chase the enemy across the Seine. He did not need much chasing. Now and again the Luftwaffe showed itself, but was swamped by our excellent air support. In fact, one was, as like as not, strafed by Thunderbolts and not Focke-Wulfs.'

Every squadron was capturing and writing off a colossal amount of kit, and 'the Guns' confidently claimed a lot of damage done, though a little touchy on the subject of civilians.

'Over the Seine at Louviers, and another traffic jam, the first for some time. And by the very beginning of September we were up to the Somme, and across it, and in the Pas de Calais, which made the V-merchants think again. At a little village called Nuncq the inevitable for once happened, and we, like the rest of the Second Army, ran out of maps. We prepared to use our silk handkerchiefs and local guide books, but the War Office came through just in time. It was curious to be racing through the Somme country, leaving in our wake such famous and bitter names as Béthune, Arras, St Pol and Vimy. We were practically out of France, enriched by the company of a large number of Maquis, who became our good and faithful friends. In Mazingarbe one joyful old man rushed up to a car, shook his liberators warmly by the hand, and shouted, "I'm so happy, I'm so happy, I'm going home, I'm going home!" "Oh, yes; and where do you live?" we said. "Caen," he replied.'

That first week in September must be the most spectacular in the history of the regiment. How many miles were covered, how many enemy killed or taken prisoner it is impossible to say. After the Seine the Somme had been crossed in a day or so, and by the end of that week the 11th were on the outskirts of Ghent. September 4 was the day Belgium was entered, and from that time on there was as much danger from flying fruit and wine bottles as from the Germans. The welcome was terrific, and all the more of a surprise after the comparatively tepid reactions of some of the French.

Then came St Nicholas and Neukirchen Waes, Waesmunster and Lokeren, where the odd Tiger made a nuisance of itself. It was not unlike an Aldwych farce: people on bicycles got chased by people in tanks, others who went out partridge shooting returned with prisoners of war, and one squadron looked so fierce that an M.P. near Toufleurs reported tham as 'an enemy column of approximately divisional strength approaching from the west'.

It seemed certain now that the war would be over in a matter of weeks, at least before Christmas. It had been won in Normandy, now it was a case of tying up the loose ends. On September 17, however, the Arnhem landings started and as the month drew to an end it became increasingly obvious that the Allies were not home yet.

At the end of October the regiment took part in the 's Hertogenbosch push, which only lasted a week. But the regiment played its part, and quite a big one, though it did not quite have the luck to be first into either Tilburg or Breda. One surprising incident occurred when a troop was conducting an O.P. on to a German gun position. The troop leader could hear the enemy speaking on his own wireless frequency, and was considerably helped by their comments on his shooting. Finally they asked permission to withdraw, which was refused (one suspects that here was a case of someone being detailed to 'die the hero's death') and the rest was silence.

The next move was (at least on paper) a short one, across the river into Holland again, where the regiment was based on Grevenbicht, Papenhoven, Obbicht and Buchten. It was getting bitterly cold now, hardly the ideal time to sit on the Juliana Canal in Roosteren or Gebroek. The cold was the worst enemy, freezing K guns and Brens and the petrol in carburettors, besides numbing hands and feet. As a result, engines had to be started up at frequent intervals during the night, which may or may not have led the Germans to believe the 11th were massing for an armoured attack. They retaliated with what the Intelligence Summaries were pleased to call a gramophone record of heavy engines, with the sound of galloping horses on the other side! At least the regiment was in a better plight than the Americans whom Hitler had caught napping in the Ardennes.

About this time the regiment set foot on the sacred soil of Germany which they had to cross to reach Jabeek, where a squadron at a time rested. And at this time, too, 'the Guns' fired their last rounds, and several thousands of them, before they finally packed up. They had done great work.

It was a change, and perhaps a relief, when orders came to move to the Breda area, where duties embraced everything from military government to assisting the line-crossers near the Maas Estuary. This was a period of flares and machine-gun fire at night, of duck shooting and boating. Every half-hour or so flying bombs or V2s would cross the river, mostly heading for Antwerp. It was a wide, open, rather bleak country, this land round the Hollandsch Diep, but in villages like Zevenbergen, Made and Hoeven British troops were always very welcome.

About the middle of March the regiment left this area, and concentrated near Weert for the Rhine crossing. There was no job for armoured cars in the immediate bridgehead, where the wonderful fighting of British and American airborne troops had paved the way for a break-out. After the flat open country of Belgium and Holland it was disturbing to be back in terrain that was often heavily wooded and where the tracks and roads were often bad—but it was a relief to see a hill again. But with memories of the *bocage* country to help them, they pushed on through Borken and Stadtlohn, Raesfeld and Ahaus. Not to have to worry about civilian damage was a great advantage. The fighting was scrappy but stiff, with bazookamen having a perpetual field day. (The bazooka, or *panzerfaust*, in passing, first made itself felt in August 1944, but was never used on such a large scale as it was in the Fatherland. White flags were seldom to be trusted, and there was of course little civilian information to help the leading Troops.

The advance went on, with odd cases of extremely fierce resistance: 'It was a peculiar advance, for you went for twenty miles without meeting any opposition, then perhaps a complete day was spent with no progress at all. So it was at Stadtlohn, and at Rheine, and to a lesser degree at Diepholz and Sulingen. The run to Rheine was the first since the previous October to compare with any of the runs in France and Belgium. Next came Bassum, Syke and up to the outskirts of Bremen. There was, of course, a continual possibility that the centre line would be cut, which, to our great surprise, and even in some cases mortification, did not happen. The rivers Weser and Aller were crossed, after a day or two's much-needed rest, and it was on to Hamburg, with the enemy becoming more and more clueless, though no less fierce. It became largely a matter of finding the right road, as there just weren't quite enough bazookamen to go round.'

What will chiefly be remembered about this most hectic month and a half? Perhaps first of all the increased number of casualties; those who fell in the hour of victory will never know how much and how successfully they contributed to that victory. On the whole, the regiment came off lightly, though many were wounded. The second outstanding thing was the vast and welcome amount of loot, or plunder, available.

The 11th Hussars were given the honour of being first

into, or rather first out of, Hamburg. Germany's second city was an impressive but terrible sight. The end had come. It had been a long journey, but a comparatively quick one. There were many things to look back on, some with pride, some with sorrow, some with laughter, but with all other emotions went a feeling of colossal relief.

It had taken almost exactly a year to beat the German armies in the field, in their own fields, moreover, as well as in France and the Low Countries. It had been a wonderful experience, and though the victory had been in many cases hard and not easily won, it had surely been cheap at the price.

Chapter 16

1945-1969

AFTER triumphant entries into Hamburg, and later Berlin, the regiment stayed in Germany as part of the British Army of the Rhine for nearly eight years. The first years of occupation were important, sometimes exciting, and certainly thought-provoking. Berlin was a strange city, and contact with the Russians, at first friendly, dwindled as they withdrew most of their troops outside the city. Various squadrons of the 11th served their time there. Elsewhere, there had been the immediate jobs of rounding up S.S. men and other dubious characters. Until about 1948 there was always the fear that many Germans who had served with Hitler's armies might not think it amiss to join the Russians. Try as one might, there was also always the German insistence that the English and the Germans should combine to attack the 'Red menace'. The blockade of Berlin and the allied air lift did much to prove that neither the Americans nor the British wanted another war, and certainly were not going to be pushed into one by the Germans.

The 11th had another tour of duty in Germany between 1962 and 1969. All of that time they were stationed at Hohne, so close to the former Belsen concentration camp that it was not amusing in the first place. Those who remembered the bestialities of the Hitler years could not easily forget; those who were new to Germany soon forgot and treated things more normally. This was excellent, because in its long stays in Germany the regiment has built up by now rapport with the Germans, remembering that they have been for many years NATO allies. The old habits of military government have given way to new habits of co-operation with West Germany. All the same, frontier patrols have always been carried out.

Perhaps the greatest hazard to soldiers' morale in Germany is boredom, despite the many benefits and even luxuries available to the modern Hussar. Great care has always been taken by the regiment to combat this boredom, not only by making soldiering itself more interesting by skilful and entertaining training, but by fostering all kinds of sport and recreation. Those who have visited the 11th Hussars since the last war, whether in Germany, the Middle East, Malaya, or at home have always found it a happy as well as an efficient regiment. As in all other modern British regiments soldiers are getting married nowadays far younger than in the old days, and a succession of Commanding Officers' and many other wives have not spared themselves in welfare activities. The fact that there has always been a family atmosphere, also noticeable to visitors to the regiment, has been a suitable reward.

Credit should go to the Commanding officers in Germany, who, from Lieut-Colonels Payne-Gallwey and Robarts in the early days to R. D. Sutton, T. A. Hall, P. M. Hamer and C. H. Robertson more recently have shown political and diplomatic wisdom as well as military efficiency. This is highly important in any station abroad, but perhaps especially in Germany.

Between July 1953 and July 1956 the regiment was in Malaya, commanded by Lieut-Colonels P. Arkwright and M. Grant-Thorold, D.S.O. Since both of these officers had served in Palestine before the last war, it was not unnatural that they had a good idea of how to cope with Communist terrorists, or 'C.T.s' as they became known. In many ways the operations in Malaya resembled closely those in Palestine. Squadrons were usually split from regimental headquarters by many miles, and operated largely on their own, and individual Troops also had much work to do separated from their own squadrons. Each squadron headquarters worked with an infantry brigade, and each Troop came under command of an infantry battalion. By this time the regiment had two Daimler armoured cars and two G.M.C. A.P.C.s to a Troop (later Saracens) and four troops to a squadron.

National Service had ended in 1960, and one result of this was that there was a danger of the standard of soldiers dropping, since in national service days there had often been some exceptionally good men who, although their stay in the Army was very limited, would become Corporals or even, in a few exceptional cases, Sergeants, before they left. On the other hand there is always much to be said for a volunteer force, and it is disrupting, to say no more, to train a man for a year or eighteen months and then lose him when he is just reaching the peak of his efficiency.

Malaya was much more interesting than Germany, however, for here the men knew and appreciated that they were really doing a worth-while job that was demonstrably seen to be worth while, even if it was galling for many ordinary soldiers and N.C.O.s not to get home on leave to Britain for two or even three years.

The regiment was fragmented because of the command system and the geographical facts of life. Thus there was usually a squadron headquarters at Kuala Lumpur for operations in the Selangor area, another at Seremban (the Negri Sembilan area), and the third at Johore (the Johore area).

Malaya gave a splendid opportunity for young officers and N.C.O.s as well as for squadron leaders, and they were greatly encouraged by having new and sometimes exciting responsibilities. Often a Troop would be stuck out in a small village under command of a Ghurka battalion, say, whose Colonel would have complete operational control. All the

11th Troop leaders in Malaya were very well trained by the time they left, if not before, and their squadron leaders too learned some new and profitable lessons. Some Troops were led by Troop sergeants from time to time. Compared with the rather flat feeling so familiar to those who have served for some years in Germany, Malaya was a decided tonic. When the time came for the regiment to leave and return to England, it was at its peak of post-war efficiency. The 11th had for the first time in Malaya used Saracens as Armoured Personnel Carriers.

Malaya was not without its disadvantages, of course. The greater part of the country is composed of dense, impenetrable jungle, clearly unsuitable for armoured car or tank work, and the climate is hot, oppressive and humid, so that after a time those who are not irritated by it tend to succumb to what Major W. K. Trotter calls 'the inevitable Malayan lethargy'. Major Trotter adds,

> 'Our main task in Malaya was to ensure the safety of all roads. This necessitated endless patrolling and bank searching, where possible ambushes might have been sprung. Our second task was food convoys. All food travelling between towns and villages had to be escorted in convoys for two reasons: first, to prevent the Terrorists ambushing food vehicles, so as to replenish their own very meagre supplies; second, to prevent Communist sympathisers dumping food at pre-arranged places beside the roads, for later collection. Our third task was V.I.P. escorts, this to ensure their safety and prevent the Terrorists from achieving a propaganda boosting kill. Finally, when time allowed them to be undertaken, jungle patrols and night ambushes.'

There was thus more than enough for the Troops to do, and each squadron probably covered something like half a million miles during its stay in Malaya; if regimental headquarters figures are added, the total must have reached nearly 2,000,000 miles in three years.

Troops conducting food convoys—which were vital to General Sir Gerald Templer's complete plan—usually spread out down the length of the convoy with the Troop leader in front setting the pace. The assault troopers from the Saracens rode 'Shot Gun' sometimes, as in an old Western movie, sitting in the front of or on top of the food lorries; more often they remained in their A.P.C.s ready for immediate dismounted action should the convoy be ambushed.

The worst problem was keeping the convoy closed up, so that if there was an attack the armoured cars were in a position to do something about it. The lorries, invariably overloaded, were always boiling over and stopping to cool off, usually in the most awkward places.

From Seremban, for example, a convoy would wind up and down the many zig-zag bends of the Kuala Pilah pass, a favourite ambush area of the Terrorists of 'No Three Fighting Platoon', which was commanded by a notorious Chinaman, one Tong Fuk Lang; this gentleman had served in the wartime 'Force 136' and had reputedly killed in a single ambush 130 Japanese on the Jelebu Pass. He had certainly been awarded the M.B.E. and marched in the Victory Parade in London after the war, but now his activities were hardly consistent with these distinctions.

Once safely through the Kuala Pilah pass, the convoy would drive through more open country, stopping periodically to offload supplies at various small villages. At the town of Kuala Pilah there was a longer halt; then on again through mainly open country with paddy fields on both sides, more villages, more halts, and then the town of Bahan. Here two Troops on detachment would join the one with the convoy, which was then broken down into small sections, all of which had to be escorted to the many rubber estates in the area. Now the terrain was ideal for ambushes, with narrow roads often passing through primary jungle and heavy foliage sweeping down frequently to the edge of the tracks. The food was delivered to the rubber estates at their guarded and wired-in enclosures.

The convoy would then reform in Bahan and return to Seremban. This is an example of one typical operation, which may sound only too easy, but somehow never turned out to be. Lorries were breaking down continually, which meant that the convoy would have to halt, offering splendid opportunities for ambush from the jungle. Also a sharp look-out on the lorries themselves had to be kept in case some food was dumped for later collection by the Terrorists. Tedious such work might be, and tiring, but it was a system that proved itself and was of great value, and, as far as the regiment was concerned, it never lost a convoy or part of it to ambushers.

V.I.P. escorts were more fun, but what the soldiers enjoyed most of all was being on Troop detachments, well

A Saracen Armoured Personnel Carrier near Bahan, Malaya in 1955. The officer in the picture is 2/Lt. T. R. Hartman.

away from the eagle eyes of the R.S.M. and S.S.M. Because these detachments were very much the Troop's own show they had an added purpose and interest, and, despite their separation, often by many miles, from their squadrons all Troops maintained an excellent standard of cleanliness and discipline.

Gungie Dua was probably the loneliest detachment, being the only one not shared with other troops, infantry such as the Gordon Highlanders, the 1/7th Gurkha Rifles, or later the 2nd Royal Welch Fusiliers, the 2/6th and 2/10th Gurkha Rifles. Each Troop had its own R.E.M.E. detachment to support it, and its own cookhouse run by A.C.C. personnel attached to the regiment.

The rubber estates were especially vulnerable, often being cut off from the nearest towns or villages by remote overhung roads that passed through the jungle. Until late in the emergency they were easy prey for subversion, threats, murder and extortion of food and money by the Terrorists. Half Troops were used for the continual patrolling needed to guard the rubber tappers, and they continually changed their routes and timings, for if a rigid routine was kept it was only too easy for the Terrorists to spring an ambush.

What sounds like rather monotonous living was interrupted from time to time by periods of intense activity and excitement when ambushes occurred and the 'C.T.' had to be dealt with and chased, or when positive information came in about the movements of 'C.T.' groups. Patrols deep into the jungle were seldom carried out because of the regiment's other commitments, but one that was, and turned out to be highly successful, was commanded by Captain P. H. Wood, and worked from Fort Iskander. In Semeli country it ran into a party of Terrorists and killed three, Captain Wood, Sergeant Stones, and Lance-Corporal Redman charging them firing as they ran. The other members of this patrol who took part in its operation were

Corporal Modelski, Lance-Corporal Mitchell, Troopers Clements, Wilkes, Bird, Clubb, Evans, Private Sneddon and Signalman Ashley. Four other terrorists got away, but were later shelled by 25-pounders from Fort Iskander (with, it was hoped, more of their comrades) but valuable documents were found on the bodies of those killed.

Major Trotter, who at that time was serving as a subaltern in 'A' Squadron, summed up of these years:

> 'My general impression of Malaya was that because of the Emergency there seemed to be a greater purpose in life than in any other station I have ever served in.'

Between 1956 and 1960 the regiment was at Carlisle, and in Northern Ireland briefly. The next two years were spent in the Middle East, at Aden, Bahrein, Sharjah and Kuwait, first under command of Lieut-Colonel J. A. N. Crankshaw, M.C.

'*Ferrets*' *of the 11th on a troop-training exercise in Aden, 1961.*

Now the regiment had four troops to a squadron, each Troop consisting of four Ferret armoured cars, and also a heavy Troop of one Ferret and two Saladins, the latter having 75 mm. guns. In Kuwait R.H.Q. were to use Land Rovers and Ferrets. The squadron at Sharjah put one Troop at the disposal of the Sultan of Muscat's forces, one at Niswa, and one at Buraimi, where frequently it had to be very watchful for snipers and mines in the sand.

At the end of June 1961 Lieut-Colonel P. D. S. Lauder was called to a conference at H.Q. Middle East in Aden where it was decided that the ruler of Kuwait must be helped in the face of possible invasion by Iraq. At the beginning of July Tactical H.Q. of the regiment and 'A' Squadron joined the Kuwaiti forces under Sheikh Saleh on the frontier, and later there were always about two squadrons in the line. Squadrons alternated, and the regiment came under 24 Brigade, commanded by Brigadier D. G. T. Horsford.

Despite the intense heat—the maximum shade temperature was often around 115–120 degrees, and of course it was far hotter inside the cars, the Kuwait operation was a great success and achieved its purpose, for the Iraquis never did invade, though the frontier was crossed briefly at times. They were soon seen off by the 11th and the Kuwaiti soldiers, with whom the regiment got on very well.

Not one 11th Hussar was *hospitalised* because of heat or sunstroke throughout the whole operation, although in the turret of a Saladin at the end of July 136 degrees Fahrenheit and 90 per cent humidity were recorded. Closed down, with the car's engine and radio on, the temperature rose to 150 degrees. A jerrican left in the sun for an hour or two made coffee which was too hot to drink.

Not the least problem was that of food and water, and the 11th much appreciated presents of large yellow lizards and melons from their Kuwaiti friends, even if these were not on their normal diet.

Sudden and violent sandstorms made communications tricky and interfered with air operations, but the Iraqui Air Force put in few serious appearances: the R.A.F. was always in control of the situation. The regiment emerged from its Kuwait experience with an increased reputation, and indeed its state of efficiency when it left the Middle East was about as high generally as when it left Malaya. Individual initiative had always been encouraged, and it is pleasant to record that when 'B' Squadron was told that it

A Ferret of the 11th in Kuwait.

must move to Kuwait the orders for the move were given out by S.S.M. Shipley.

Brigadier Horsford said after the Kuwait operation: 'It has been the most wonderful training in co-operation for all three arms of the Service. The troops put up with the conditions extremely well and morale has been absolutely sky-high the whole time.'

Air Chief Marshal Sir Charles Elworthy, Commander-in-Chief Middle East, sent a message to Colonel Lauder on July 10, 1961:

> 'You have done a really first class job in your vital position out on a limb and to add to your success the excellent relationship that you have established with Sheik Saleh and his force has been most important politically, quite apart from the obvious military value. . . . Thank you all for the fine job you have done.'

This was as pleasing a commendation to get as had been that from Major-General R. N. Anderson, G.O.C. 17 Gurkha Division, on July 16, 1956.

> 'Yours has been the vital task of keeping the roads of South Malaya safe from the enemy; this you have done right well, and in so doing earned for yourself a reputation second to none for both operational efficiency and your cheerfulness and good humour in the face of difficulty. On the occasions when you have had the opportunity to close with the enemy you have always displayed that "élan" for which the 11th Hussars are renowned throughout the Army.'

When the regiment returned to Tidworth from Aden at the end of 1961 it had to convert itself during the next six months from armoured cars to tanks. This was a most drastic step for a regiment which had become famous for its armoured car work over some thirty-three years and considered itself, not unjustly, the world's expert on reconnaissance. Now it was expected to reach the same high standard in a completely new job in a very short time. Colonel Lauder made it clear from the start that this would be achieved. The next six months were among the most hectic that the regiment has ever known, as each officer and man had to be taught how to use a completely new set of equipment, all tactical thought had to be revised if not reversed, and there were a hundred and one other things to be done in individual and collective training, administration, and servicing. A high standard of all-round training was achieved, largely due to the careful thought and hard work of Major George Hodgkinson, the second-in-command, and of his instructors. All the same, when the time came for the regiment to move to Luneburg Heath in the summer of 1962 it still could not count itself fully trained.

As part of 7th Armoured Brigade, in 1st Division, which

Aden, 1961. 11th Hussars on the Dhala road at the foot of the Khoreita pass.

had inherited the traditions, spirit and the Jerboa of 7th Armoured Division, the 11th Hussars provided patrols from the Reconnaissance Troop along the miles of wire and minefields separating West from East Germany, besides taking part in regular tank training and manoeuvres. The tanks used were Centurions, and later some Conquerors, during these years in Germany between 1962 and 1965. In 1964 Colonel Lauder gave up command after a most successful and efficient two and a half years of command; he was succeeded by Lieut-Colonel R. D. Sutton, the first officer to command the regiment whose father had also commanded it. Under Dick Sutton's command the regiment's efficiency as a tank regiment continued to improve; but perhaps the highlight of his time was the presentation of a new Guidon to the regiment in 1965, on the occasion of the 250th birthday of the 11th Hussars, by H.M. Queen Elizabeth the Queen Mother. This was a splendid occasion enjoyed by all, and many Old Comrades came out from England to take part in the festivities. Colonel Sutton presented not only an immaculate parade, but also delivered a short speech in reply to the Queen Mother's words that hit exactly the right note. He claimed without complete conviction that he had made it up that morning in his bath.

Not long after this occasion Colonel Sutton took a large detachment of the regiment to Coburg in Bavaria, the home town of Prince Albert, and this long overdue visit was a resounding success. Its conclusion was marred by tragedy for on the way back to Hohne Colonel Sutton's staff car had a serious accident, hitting a German lorry head on. His driver was killed instantly, and Dick Sutton died in hospital two days later, his wife being seriously injured, but now happily recovered.

Major T. A. Hall now took over the regiment for six months, with Major S. D. Bolton as second-in-command. Lieut-Colonel Hall's period of command was notable because during it the regiment was given the great honour of being told that it would be the first to get Chieftain tanks, which, although their prototypes had earned rather a bad name, turned out to be absolutely first-class when various troubles had been corrected. They were certainly a great advance on the Centurion, good as that was. The Conqueror had never been any sort of success.

This translation to Chieftains again meant a great deal of hard work and new training, and Majors Bolton, Hamer and Sergeant Gormley, B.E.M., deserve particular praise. But in fact everyone in the regiment did their bit, and one senior R.T.R. officer, who might well have looked at a

Stables—the 11th on parade in Germany.

cavalry regiment with some scepticism, said later that the 11th Hussars had proved themselves the most efficient tank regiment in Germany at this time. This testimonial, given in private conversation, was completely unsolicited and very heart-warming.

Colonel Tom Hall was succeeded by Lieut-Colonel Peter Hamer, who kept the regiment in a high state of efficiency and was awarded a well-earned O.B.E. in 1969—the first in the regiment since Colonel Arkwright had been so decorated in Malaya. In October 1968 Colonel Hamer handed over to Lieut-Colonel Clive Robertson, who had the distinction, albeit a sad one, of being the last commanding officer of the 11th Hussars. But his calibre has been recognised in that he also becomes the first commanding officer of the Royal Hussars, the name of the regiment formed by the amalgamation of the 10th and 11th Hussars. Amalgamations of old and famous regiments are always regrettable, but both regiments have approached this one in an admirable spirit of co-operation and forward-looking enterprise. Both have been helped by the guidance and example set by their Colonels, Major-General David Dawnay, C.B., D.S.O., and Colonel Sir John Lawson, Bt., D.S.O., M.C., as well as by their own commanding officers.

Today's soldier has to be a great deal more technical than his ancestors, even his predecessors of thirty or forty years ago. With equipment in a tank regiment valued at over £10 million in vehicles alone, very likely, and an annual cost of running a regiment of forty-six officers and 571 other ranks standing at nearly £4 million, one has to know one's stuff and look after one's equipment very carefully. Many old soldiers would hardly recognise the form of modern soldiering. As one officer puts it:

> 'Every man must be much more self-reliant than ever before; he has complicated equipment to operate and maintain, and he has to be able to operate by day and night for long periods without a break, wearing gas

masks and protective clothing as often as not. Distances of engagement are now in the order of 3,000 yards where they were only in hundreds of yards a few years ago, troops and squadrons are further apart and must be able to operate for lengthy periods on their own; the soldier of today must be more intelligent, more mechanically minded and more independent than ever before.'

A slightly different aspect of life was put amusingly in the Regimental Journal of December 1967, under the title, 'You've Never Had It So Good.' The author writes:

'Gone now are the days when "smoke-break" meant a tipped Woodbine shared with two mates and "fall-out" was good for you. Do you remember the days when a Sabot was on the foot of a mademoiselle from Armentières and Balaclava was worn and not celebrated? As old soldiers throughout the world reminisce furiously, vast changes are occurring in every Army Camp from Hohne to Hong Kong. Walk in the main gate, "Welcome, Speed Limit 50 k.m." On leaving "Thank You and Safe Journey." Such a change from a Demob Suit (some of which are still worn), a Railway Warrant and the option to buy your greatcoat, cheap rate for ex-service men.

'Enter any barrack room and wonder at the Psychedelic decoration which confronts you. Foam mattresses, clean white sheets and grey steel lockers, which appear to have recently been removed from a Nuclear Submarine. Again a fantastic difference from whitewashed Nissen Huts, palliases filled with a mixture of straw and waste paper, hairy blankets and wooden lockers knocked up by a carpenter in the throes of a trip on LSD.'

A soldier going on leave nowadays, continues the writer:

'will get up an hour before the bus leaves for the airport, throw on his Carnaby Street creations and disappear to Majorca for three weeks ("at 10 pounds a week clear I can afford to"). Oh for the days of two weeks in the Union Jack Club. Night life; first week Soho, second week Bow Street Police Station. . . .

> 'It's true that "you've never had it so good" but in some cases "They haven't had it at all." '

The old order changeth, yielding place to new, but the 11th is still a happy and efficient regiment, and there is every sign that the new amalgamated regiment will be the same, since the 10th and 11th have known each other for many years. The last annual inspection report in Germany in December 1968 was by Brigadier R. C. Ford, M.B.E., Commander 7th Armoured Brigade, and he had this to say about the regiment, and its return early in 1969 to England.

> 'As a result of this move the Regiment, the first in the British Army to be equipped with Chieftain tanks, is to revert to Centurions.
>
> 'In the light of these factors, it would have been understandable if 1968 had proved to be a year of running down and running out. Nothing in fact could have been further from the truth. The Regiment has continued to function in top gear and the very high standards that anyone who knows the Regiment has come to take for granted have been maintained.
>
> 'Throughout my inspection I was impressed by the cheerful and professional reaction of all ranks to my questions. A superficial attitude of cheerful light-heartedness conceals a fierce pride in the Regiment and a belief that only the highest standards are acceptable.
>
> 'It has been a privilege for me to have this famous Regiment, with its long and distinguished history, under my command. They leave 7th Armoured Brigade on the highest possible note and take to the Royal Hussars quality which will guarantee the sure foundation of the new Regiment.'

The 11th Hussars in the last war were longer in contact with the enemy than any other regiment in the British Army, won more battle honours than any other cavalry or Royal Tank regiment, and had, surprisingly in view of their continual fighting, almost the same number of fatal casualties as in the Great War. Brigadier Clarke wrote:

> 'Yet from it all they emerged with their gaiety, their sense of humour and their stark professional efficiency quite unimpaired. In the survival of these qualities there may be found, perhaps, the true secret of a great success.'

It is good to think that these qualities still abound today. It is not so good to be reminded constantly by the Government that, having already done its best to destroy the

A far cry from the horses of the Crimean era: the Chieftain tank.

Territorial Army, it is equally unscrupulous about abolishing or amalgamating famous regiments that have much to offer in peacetime as in war. The 11th have been more fortunate than some other regiments with equally cherished traditions and proven abilities, but it is, to say no more, a great pity that those in political authority find it necessary to destroy or condense something which they see to be functioning perfectly.

The position cannot be put better than it has been already by Colonel Sir John Lawson, and on this note this book should end.

> 'I am sure that all 11th Hussars will agree that these are unpredictable days for their Regiment and, indeed, for the Army as a whole. Assurances that are given on one day are cast aside on the next; standards of honesty in speech are at their lowest ebb. This leads me to consider that the spirit of Regiments, such as ours and the 10th Hussars, which, has caused them to serve their country so outstandingly through so many years of history, is more important today than it has ever been. Let us hope and, indeed, pray that we may now be left alone in our endeavour to create something new out of something old and to be determined that tradition, comradeship and high standards are very worth-while things.'

If the regiment remains true and faithful to its past and present standards, there should be little fear that the new combined regiment, the Royal Hussars, will also be a great success.

REGIMENTAL MUSIC

Note by Bandmaster, May 1968

THE composer of the Regimental Slow March COBURG is anonymous although it is supposed traditionally to be by an aunt of Queen Victoria. This is properly known in Germany as Prinz Josia's March. (He being a General of the Duchy of Saxe-Coburg.)

The Regimental Quick March is MOSES IN EGYPT, an extract from the Oratorio of the same name by Rossini. Cavalry Regiments, never marching past on foot, had no requirement for a Quick March and used only a Slow March. The reason for their choice is obscure; it may perhaps be the pre-war association with Egypt. Since the end of World War II all Regiments of the British army were required to have a Quick March.

LILLI MARLENE, a German 'pop' tune was used by the Afrika Corps and the 8th army. It is little used now except at reunions.

OLD GREY MARE was adopted by the 11th Hussars after the 1914–18 war, with words sung by the soldiers—no doubt bawdy—as an unofficial signature tune.

11TH HUSSARS BAND

THE first mention of any sort of band is attached to the date May 20, 1785, when the 11th Light Dragoons apparently had a 'Band of Music consisting of ten, four of whom are enlisted as Dragoons.'

According to the XI Hussar Journal of April 1914:

> 'The first record of attested soldiers being employed as musicians in a Regimental Band is that of the Coldstream Guards, for whom in 1785 the Duke of York enlisted twelve musicians in Hanover, one of whom received the appointment of "Music Major".'

In 1822 Regimental Bands were fixed at ten musicians, 'not including black men and boys.' In 1823 this number was increased to fourteen. Since 1684 the Regiments of Foot Guards had been authorised to employ twelve oboes—presumably between them—and nine other regiments followed this example. The oboe may thus be regarded as the starting point for regimental bands.

Coburg

The Eleventh Hussars (Prince Albert's Own)

1715	Raised for King George I by Brigadier Philip Honywood at Chelmsford, and known as 'Honywood's Dragoons'. 'The Fifteen' and Battle of Preston.
1716—44	Home service. In 1732 Regiment becomes 'Lord Mark Kerr's Dragoons'.
1745	'The Forty-Five.' Clifton Moor (last notable fighting on English soil), Culloden.
1746—59	Home service. In 1752 Regiment becomes 'Ancram's Dragoons'. From 1755 the number '11' is constantly associated with them. In 1758 the Light Troop takes part in raids on French coast.
1760—63	In Germany during Seven Years' War. Warburg and other battles.
1763—92	Home service. In 1775 Regiment becomes 'Johnston's Dragoons', but more usually known as '11th Light Dragoons', a title that becomes official in 1783.
1795—99	War with France. Regiment in Flanders. Sieges of Valenciennes and Dunkirk. Battles round Le Cateau and in Holland.
1800—11	Home and Irish service. In 1800—01 'C' Squadron is part of Abercromby's Egyptian expedition. Battle of Alexandria.
1811—14	The Peninsular War. Regiment in many actions. Nicknamed 'Cherrypickers' because a detachment is traditionally surprised dismounted by French cavalry in a cherry orchard. Battle of Salamanca.
1815—18	Waterloo. Occupation of France.
1819—38	India. Siege of Bhurtpore.

1838—54	Home service.
1854—56	The Crimean War. Battles of the Alma, Balaclava, Inkerman, Sebastopol.
1857—65	Home and Irish service.
1866—77	India.
1878—1903	Home and Irish service, South Africa, India (Frontier 1897–98), Egypt. Various detachments and drafts sent on active service elsewhere, notably during Boer War.
1903—07	Ireland.
1908—14	England.
1914—18	The Great War. In France and Flanders throughout, often fighting dismounted in the trenches. Néry, Messines, many other battles.
1918—26	The Rhineland, Egypt, India.
1926—34	Home service. Converted to armoured cars in 1928 after 213 years as horsed cavalry.
1934—39	Egypt, Palestine. Arab revolts.
1939—45	The Second World War. North Africa. Regiment serves in the Desert campaigns throughout, except for brief interval in Iraq and Persia. Italy 1945. Normandy-Berlin 1944–45.
1945—53	Germany, in B.A.O.R.
1953—56	Malaya.
1956—60	Home service.
1960—61	Aden, Persian Gulf, Kuwait, Trucial States, Aden.
1962	England, conversion to tanks completed.
1962—69	Germany, in B.A.O.R.
1969	England. Amalgamation with 10th (Prince of Wales's Own) Royal Hussars on Balaclava Day, October 25.